"IF YOU DON'T LIKE THE WEATHER..."

Stories of Texas Weather

"IF YOU DON'T LIKE THE WEATHER..."

Stories of Texas Weather
by John Edward Weems

★

TexasMonthlyPress

Texas Monthly Press, Inc.
P.O. Box 1569
Austin, Texas 78767

A B C D E F G H

Library of Congress Cataloging-in-Publication Data

Weems, John Edward.
If you don't like the weather.

Bibliography: p.
1. Texas—Climate. 2. Weather—Folklore. I. Title.
QC984.T4W44 1986 551.69764 85-28821
ISBN 0-932012-84-1

Design by Lonestar Studio
Illustrations by Valerie Kerr

For my friend Lynn Roberts Grice,
And for my aunt and friend, Lucy Hill

CONTENTS

Also by John Edward Weems

The Tornado
Death Song
To Conquer a Peace
Dream of Empire
Men Without Countries
Peary
Race for the Pole
The Fate of the Maine
A Weekend in September

EDITOR:
A Texas Christmas

CONTRIBUTOR AND EDITOR:
Black Art in Houston
U.S.S. Ashtabula*: Her History*

CONTRIBUTOR:
Six Spanish Missions in Texas
Texas Vistas
Growing up in Texas
The Schiwetz Legacy
A Vanishing America

Dictionary of American Biography
Collier's Encyclopedia
The Handbook of Texas

ACKNOWLEDGMENTS

More individuals than one might expect helped with this brief book, but none more than Lynn Roberts Grice of Houston, who researched sources in the Texas Medical Center Library for weather- and health-related topics. She gave me hundreds of pages of notes, enough to make a popularly written book on that subject alone at some future time. Margaret Cooper (Maggie) Neighbors of San Antonio sent me copies of old family papers and clippings dealing with the two hurricanes that struck Indianola. Leo Lyons of Waco let me borrow his library and files on Texas weather.

A complete list of individuals who helped: David Ballew and Mary Ballew, Gholson; John Banta and Leo Lyons, Waco; Liz Bardsley, Leslie Carpenter, Anita Walker Howard, Lavonne Ishman, John H. Kyle, Scott Lubeck, Lucile C. Lyda, Anne Norman, Pat Folmar Robinson, Arnold Rosenfeld, Fred Sackett, and Barbara Waldrop, Austin; Betsy Colquitt, Fort Worth; David Grice, Lynn Roberts Grice, Eric Mair, and Jacqueline Simon, Houston; Lucy Hill, Moody; Charlie Hooks, Spring; Lloyd Lyman, College Station; Jane McMillin, Lampasas; Margaret Cooper Neighbors, San Antonio; Judyth Rigler, McAllen; Barbara Rodriguez, Blanco; Jane Weems, Waco; and Carol Wortham, Beaumont.

PART ONE

1. WEATHERED RECORDS

My original interest in Texas weather can probably be traced to an old frame barn that no longer stands atop a Central Texas hill eight miles east of Temple, in Bell County.

When I knew the barn, in its waning years, it was leaning further and further northwestward, toward dilapidation. It had never been painted, in keeping with pioneer Texas tradition. But even in its windblown antiquity it served as satisfactory shelter for the animals on my grandfather's 200-acre farm. It also contained, handwritten in pencil on interior walls, an almost-daily record of Central Texas weather over many years.

These weather notations represented one of only two kinds of writing I ever knew my grandfather to sit still for. He also wrote a few close and favored individuals post cards, which in those days cost a penny. About once a month his grown-and-gone children, my father included, would receive brief, penciled messages that usually declared things on the farm were "alright" (his spelling), even though in my memory they frequently were not all right—at least not financially. Many of the poorest years I remember were known then as "bad times," a description that later evolved into the Great Depression.

Invariably, as I recall, my grandfather would add to his brief message at least one sentence about the local weather. To him, discussion of this topic never represented triteness. Whatever remarks he made on the weather reflected, matter-of-factly, its predominant importance—then, now, and probably forever—to farmers and stockmen. Whenever he mailed a post card in July or August that said, "No rain in five weeks," the recipient, invariably possessing prior knowledge of the farm, knew that things could not really be "alright."

The stock tank would be dwindling or even completely dry, requiring that the farm animals be driven along dusty lanes to somebody else's vanishing water hole, if permission could be obtained, or (likelier) that most of the animals be sacrificed to stingy buyers and the remaining milch cow and pair of mules be watered the same way that the human inhabitants consistently received their replenishment: from barrels, filled daily, bucket by bucket, at a neighbor's deep well half a mile away, and hauled home on a mule-drawn sledge.

The only difference thereafter between thirst-slaking beast and man or woman lay in the containers used and the manner of intake. The animals would slake themselves during long stands at a newly filled trough of water, taking it in with continual swirling noises. The humans would get their refreshment from a canvas-topped barrel set in the shade of a lone hackberry tree that stood in an otherwise barren front yard. They would first remove the canvas, then, using a metal dipper that usually hung from a nail in the tree, dip into the cool barrel and drink water that, on the hottest summer days, tasted as though it could have been scooped from a stream high in the Rockies. After that they would carefully replace the canvas cover, to keep leaves and asps and other unwanted trash from polluting the water.

During droughts some neighbors would pile their accumulation of dirty clothes, linen, and bed covering in a wagon and rumble away early in the morning to miles-distant Little River, which often retained a few water holes even in dry weather. There they would wash and dry their clothes before returning home that night.

The phrase "No rain in five weeks" scribbled on my grandfather's post card would send another message to the recipient, too —one silently telling of vegetation turned brittle-brown, resembling plant life after a hard freeze, and of deep cracks cutting through depleted soil, forming chapped earthen lips waiting for relief that only plentiful moisture could bring.

It really wasn't an easy life at all on that farm. I remember that an uncle who grew up there (as the youngest of eight children) used to claim, with only a half-laugh, "It's the only place in the United States where *India* sends CARE packages." During those dry days, my grandfather, who had the old-time farmer's knack for predicting rain, or lack of it, would often be seen squinting skyward, looking for any hopeful sign but seeing only cloudless skies or maybe some low-floating morning puffs that he knew would soon burn away and leave absolutely nothing visible between his farm and the withering sun.

I see him, forty years after drought or rain ceased to have any earthly meaning for him, peering upward and around the horizon for obvious weather signs—as he did several times each day. That was what he referred to as "watching the weather." He didn't live long enough to watch the weathercasts on television, and considering the babble from some of those show-biz people, it's just as well for his continued resting in peace now. I never saw him make one of his weather entries on a barn wall, and if others did they never mentioned it to me. Instead, those briefly written records would simply pop up in previously blank spaces, like new sprinkles on a windowpane.

Probably my grandfather never wrote weather notations except when he was alone. Whenever he was with another person he usually talked constantly, in a loud drawl and with words that, I realize now, echoed early Texas. To him a porch was always a "gallery." Thunder and lightning "'way off yonder to the west" was a good sign (assuming the farm needed rain, which was often a valid assumption) because weather usually came from that direction. The "young 'uns" stupidly exposing their swimsuit-clad bodies for summer tanning in the 1930s and '40s, when I was one of those imbeciles, didn't risk sunburning; they invited "blistering." And so on.

I wish I could remember the exact wording of his weather notations, but I can't. They would always be dated and would look something like this: "Nov 27 1929 [or, as it was written occasionally, 11/27/29]. First freeze of season. Temperature 30°." That's only a fictionalized sample; I don't know when the first freeze occurred at that location in the autumn of 1929. Whenever it was, the notation in the old barn would have recorded his own thermometer reading (he tended to trust himself above everyone else) and would not have been copied from whatever weather information might have appeared in the local *Daily Telegram,* which he read thoroughly during a midday work halt.

One entry I do remember rather well (because my father pointed it out to me) was of an August 17, 1915, Galveston hurricane. That bit of information *must* have come from the *Daily Telegram* or some other secondary source—but not from radio, which remained absent from the farm until the arrival there, in 1938, of electricity, courtesy of President Franklin Roosevelt's Rural Electrification Administration. I recall seeing these four words, "big storm on Gulf," followed by the amount of rain that eventually fell locally from those same clouds. The coastal death toll reached 275, but my grandfather did not record that, possibly because he did not have a chance to count the casualties himself.

My grandfather's ability to foretell the weather, and his age, led me to look upon him as occupying about the same status as the venerable seers described in the Bible, although he wasn't really that old, of course, even if I did think so at the time. Poor predictions never did bring forth any profanity that I knew of from him, nor did any other suffering caused by unfavorable weather; but I do remember hearing him swear, many times, at a pair of recalcitrant mules. I suppose he thought he could do something about their contrariness, whether or not he was overestimating himself. The weather was a different matter. All he could do was watch it quietly and try to foretell changes, even if they might not be of any benefit.

His attitude was like that of the Englishman who is said to have observed, years ago, "There is really no such thing as bad weather, only different kinds of good weather."

I envy anyone who can truly believe that, especially if he or she has endured West Texas drought and dust storms and unmitigated summer sunshine, as I did in my youth. It is true that the atmosphere, whatever state it happens to be in, is one thing that keeps us alive on our planet. Thus any momentary sort of atmospheric condition—hot or cold, dry or wet, calm or stormy—is decidedly better than none at all. But I believe in the term *bad weather,* even if my belief might differ from the norm. As a native Texan—and not one from the wetter eastern part of the state—I rarely call a slow, soaking rain "bad weather." In fact, I

"IF YOU DON'T LIKE THE WEATHER..."

A windmill in Medina County. 1917. Windmills opened new areas of Texas to settlement, as immigrants no longer had to rely on a surface water supply.

still enjoy walking in it, which might create doubt among some people about my sanity.

Even though my grandfather never addressed the weather the way he did his mules, it seems to me that the weather usually treated him much worse than those animals did. Admittedly, weather never put him in a hospital. He didn't spend a day of his life in one of those places, and to my knowledge he never was sick at home, either, until his final days, at the age of eighty-two. But I remember from my visits to the farm a few years of flooded cotton fields and many more years of thirsty cornstalks, stunted and yellowed from lack of rain. Looking back on those summer visits, I don't recall any extended periods when rainfall came in balanced amounts, although there must have been a few. Consequently, I never remember that my grandparents had enough money; yet they never received welfare checks (which were mostly unavailable then), and President Roosevelt's aid to farmers never trickled down to them in any quantity.

It hadn't been that way, say kinfolk old enough to recall better days. Until 1907, when my grandparents bought the farm, my grandfather had rented, farmed, and lived on fifty acres "across the creek" (Little Elm, also in Bell County) from the place where the barn full of weather records stood, and he had somehow flourished financially—sufficiently to make a down payment on the new farm and later to pay it off. But extended periods of illness hit the family, with the exception of the male head of it: measles, typhoid, influenza. Special nurses had to be hired and doctors summoned from Temple occasionally, so that fifty acres of the new farm were sold to pay medical debts. Then about the time I was born, a drought came along, ending chances of economic recovery, which thereafter never did ensue on that farm because of the Great Depression.

My grandfather died at the close of World War II while I was thousands of miles away in the Pacific. Months later, when I returned, I visited the barn, then in the care of an uncle of mine, and read over many of the old weather entries, plus some new ones from more recent years. I noticed that the same wind and weather of my grandfather's time had continued to batter the barn. It was leaning even more decidedly toward the north, away from the prevailing wind on that hill.

Not long afterward I heard that the barn had collapsed. My first thought was of all those weathered records. Someone should have transcribed them, I thought to myself; I was determined to go back and retrieve whatever notable entries I could find in the wreckage. Varnish would keep them legible indefinitely.

Passage of time took care of my resolution. My uncle built a pig pen with the debris. That, too, later disintegrated in the wind—or maybe it caught fire and burned during a summer drought. I have forgotten what happened. Anyway, the records have gone to dust, as has my grandfather. But on a summer night a southerly breeze still moans on that hill, I'm sure, and drought, rain, sunshine, hail, and snow surely visit the place still, even if no one around there now observes and records these happenings.

The old weather prophet and his barn full of observations exist today only in my memory. But the various atmospheric conditions he experienced and strove to foresee are as operative as ever, even if humankind has managed to modify some of the effects—by, for instance, laying a pipeline that now carries fresh water directly to the old farm. And certainly the weather is often just as exasperating, but still as interesting, as ever. At least it should merit the attention of any Texas resident who breathes Texas air and drinks Texas water.

This book represents no meteorological study, obviously, but it is an attempt to say something about the state's weather as it has affected people here, or sometimes as they have only envisioned those effects. Weatherlore and folklore, I call it. There's no claim at all to completeness; that would require volumes. Every reader will probably have several weather stories more momentous than the ones told here. But I do hope this book represents a weather sampling that will provide some entertainment and maybe a little enlightenment.

2. WITHERED RECORDS

Texas is long on weather, both good and bad, but short on written records of it. A few Anglo-American accounts exist as far back as the early 1800s, mostly in diaries and other personal records. Before that, French- and Spanish-speaking people left occasional references, but not until the waning of the nineteenth century did the state have enough observers scattered around to keep an adequate journal on weather everywhere, including the prickly reaches of far West Texas.

The world's first written observation of any detail antedated that by an unknown number of multicenturies. It appears in the second chapter of Genesis. Dry land had emerged on our planet, it is reported there, and with it had come grass, herbs, fruit trees, and the like. But there had been no rain, which really wasn't so bad then, because there were no men or women around to till the soil, anyway. Then "there went up a mist from the earth, and watered the whole face of the ground."

Following that there appeared on earth the gardener named Adam, then came the Eve affair, and, eventually, banishment for both of them from Eden. Never again thereafter would cultivators of the soil enjoy perfect weather from day to day—a bad fortune that seemed to peak during my grandfather's farming days. Compounding that, thorns and thistles (as the record says) and probably other pests like mesquite trees, tumbleweeds, and unwanted Johnson grass commenced thriving outside Eden. In time, Adam and Eve might have learned to bale the Johnson grass as hay for their cattle, and they could use mesquite beans as fodder during droughts. But the tumbleweeds must have been an utter nuisance. Breaking off from their roots in dry, dusty autumns, they would have piled up along whatever fence rows had been built. Some of them eventually blew as far as West Texas, where I saw them in my early youth, during Dust Bowl days of the 1930s, bunched up against the upper strands of barbed-wire fences whose lower extremities had been buried in sand. Now, of course, I know about the availability of weather and climatic records older than the one referred to in Genesis, but it takes more of a scientist than I am to observe and write about them. I'm incapable of visualizing a time when land nearest the North Pole abounded in sequoia trees, alligators, lemurs, and ancient horses, as has been claimed.

Some scientists who have written about these things say that one reason for the world's warmth then was the existence of numerous active volcanoes spewing carbon dioxide gas into the atmosphere. This absorbed heat radiation and prevented its reflection back into space, thus keeping it here. (But other scientists whose works I have read say that volcanoes spewing any pollution into the atmosphere would make things colder.)

During those years of extreme heat Presidio, Texas, must have been the worst place in the world for a summer vacation. It's no bargain even now, not even in a comparative ice age that scientists say began a million or more years ago, give or take a few eons, before the emergence of humankind. Today, cool still bypasses Presidio in the warmer months, even if other locations have registered higher maximum temperatures in recent Texas history.

Presidio's high has hit only 117 (as I write this)—at least twice, in 1957 and 1960—but this and other maximum temperatures reported by scorched observers there leave me wondering whether, even in climes of other times, polar bears and other furry specimens ever might have inhabited *that* area. On the other hand, I can't imagine West Texas ever having had a continuing surplus of water; yet in the Permian period a sea indeed covered most of that region.

Since writing my first book in 1957, about a hurricane, I have gathered some more recent weather and climate information that is more easily understood. In Texas, plentiful indications of storms and droughts are recorded, though not in detail, and *that* sounds more like the Texas we all know today.

A storm shipwrecked and cast ashore the first European visitors to the Texas interior, an expedition led by Pánfilo de Narváez and including Cabeza de Vaca, in 1528. Some historians say these unlucky Spaniards landed at Galveston Island; others say it was elsewhere. Wherever it was, some Spaniard named the island Malhado—"misfortune"—with reason that continued to be logical. Winter proved so severe that only fifteen of the eighty stranded Spaniards survived it.

Twelve years or so later, another Spanish explorer, Francisco Vázquez de Coronado, and his men were wandering around the drought-plagued Llano Estacado (Staked Plain, in eastern New Mexico and the western Texas Panhandle), nearly dying of thirst, when friendly Indians guided them to a stream that Coronado (according to a story) immediately named Brazos de Dios (Arms of God). This became known later simply as the Brazos, at 840 miles the longest river in Texas. If the water tasted then the way it did when I sampled it long after Coronado, he and his men were indeed thirsty.

The French explorer René-Robert Cavelier, Sieur de la Salle, came across that same Brazos and, according to another story,

A tornado over the state capitol in Austin. May 4, 1922.

named it the Maligne. H. E. Bolton, the historian, believed the Brazos to have been the site of La Salle's assassination by malcontents in his expedition. Whatever the exact location of that site, it can be speculated, at least in jest, that the vicissitudes of early Texas, especially the tricky weather, could have resulted in the slaying of La Salle.

For some proof, look at *Joutel's Journal of La Salle's Last Voyage, 1684–7,* as it appeared in Henry Reed Stile's 1906 edition, and particularly at Joutel's description of what was possibly the first celebration of Christmas in Texas. The La Salle expedition had suffered shipwreck, loss of stores, and other misfortunes. Joutel wrote, "Monsieur *de la Sale* being recover'd from his Indisposition, Preparations were again made for his Journey; but we first kept the Christmas *Holy-Days.* The Midnight Mass was sung, and on Twelve-Day, we cry'd *The King drinks (according to the custom of* France) Tho' we had only water." Frenchmen without wine surely would have been rebellious, to La Salle's subsequent sorrow.

At Galveston in 1818 weather drove aground four ships belonging to the pirate Jean Lafitte. Little more than a decade after that, the founder of an American colony in what was then the Mexican state of Texas, Stephen F. Austin, was riding with a group of horsemen across the flat coastal plain near Galveston Bay when the men came upon a sailing vessel rotting on dry land.

"Some day," Austin is said to have predicted, "the elements which did that will sweep over this coast again." Ensuing storms have shown Austin's ability as a hurricane forecaster, but in another area he failed, himself a victim of Texas storminess. Years later, weakened by overwork, he developed pneumonia and died from the cold blasts of a norther that swept through his flimsy Columbia, Texas, office building in December 1836. Ironically, his father, Moses Austin, had suffered similarly from cold north winds in 1820/21 and died while preparing for the colonization of Texas, which his son then directed. For some reason, possibly to avoid discouraging immigrants, Stephen Austin never detailed in his writings this wintry Texas phenomenon.

Austin and his colonists came to know most intimately the areas just inland from the Texas coast, to a distance of two hundred miles or so. In those days the deep interior of Texas, as it appears on maps now, was largely unknown. Early travelers into Texas observed a landscape that changed gradually from east to west. A forested eastern area received thirty inches or more of rainfall annually. Farther down the coast and inland from it, a rolling central section had less rain and fewer trees except in the river bottoms, but more grassland. On westward, toward the Rio Grande, lay a sun-baked country of scrubby trees and grass, with yearly rainfalls of ten inches or less.

Far northward lay a region that Austin's colonists never saw: a high plains area, now the Texas Panhandle—treeless flatland with a fair amount of rainfall, grass, and winters remarkably cold for so moderate a latitude. Southwest of the plains rose barren mountains, in the vicinity of today's Big Bend country: ninety peaks more than a mile high, including Guadalupe, 8,751 feet. Eastward from there, toward present Central Texas and the city of Austin, lay a wooded hill country. In Stephen Austin's day Indians inhabited all this inland area—high plains, mountains, hill country.

The coastal area alone, however, proved sufficient to attract many Anglo-Americans looking for new opportunity, and most of them found the weather agreeable. Austin encouraged their immigration in letters that described the climate as "delightful" and "compared by some to Italy." Farmers who understood grape culture and winemaking could produce an important commercial produce, he said, and he boasted of fine crops already produced: sugar, corn, wheat, and "all kinds of [other] vegetables. . . . Pasturage is inexhaustible and green winter and summer." The last statement surely would have appealed to farmers who always faced harsh winter weather elsewhere. Asa Brigham, an immigrant farmer from Massachusetts, wrote his kinfolk that he might return home to visit someday, but not to spend the winter. "Texas has more natural advantages and in fact is superior to any other part of North America," he told them.

Thomas Decrow, an immigrant from Maine, wrote home a colorful description of relaxed life on Matagorda Bay, where he lived with a young wife and baby. He had built his four-room residence himself, with breeze-catching porches on two sides, about fifty yards from the water. Even as he wrote, he watched his hundred cattle grazing half a mile inland, in plentiful year-round pasturage. Animals and fowl, both domestic and wild (including deer), and his garden provided most of the food. He also had fish and oysters available from the bay nearby.

But in those precarious years such an easy life could be quickly reversed. One reason was the weather, which was not always indicative of the mild climate described by the man from Maine. An immediate concern around Matagorda Bay would have been Gulf storms and "overflows," as the occasional windswept tidal inundations came to be called. Other sudden weather changes typical of today's Texas occurred, even in the ordinarily temperate area around the early-settled Gulf Coast, and early immigrants were not prepared for some of them. Stephen Austin was not, and maybe his failure to warn newcomers proved fatal for a few of them, too.

Most of the early Anglo-Americans lived in hurriedly built log cabins with floors of dirt or clay or, at best, split logs with the flat sides turned up. Cutting through the middle of many such residences were "dog runs"—actually hallways open to the weather to catch cooling summer breezes. But with no doors available to close off this windy space, fierce north winds, though comparatively infrequent so far south, proved to be very uncomfortable.

The immigrants' clothing also indicated their expectation of finding warm or moderate weather. Apparel originally brought from the United States eventually wore out, and most settlers then had to make their own. None of it required the time and expense involved in making clothes for northern winters.

The settlers were able to provide food with their own hands: wild game or beef or pork; coarse bread made from cornmeal; sweet potatoes but, with a few exceptions, rarely any other vegetables; and strong coffee, usually black, and sometimes brewed from ground corn or some other penurious substitute. Many visitors of that day talked about a remarkable absence of fresh milk and butter from most tables, despite the presence of cattle. Occasionally buttermilk would be available.

The paradise described by those early immigrants could quickly turn to hell be-

"IF YOU DON'T LIKE THE WEATHER..."

A flood in Bandera County. Spring 1978. Rain-swollen creeks and rivers in the Texas Hill Country can rise to flood stage swiftly and sometimes without warning.

cause of the weather. Fevers raged in river bottoms and other stagnant areas along the low Gulf Coast, especially during summers, when temperature and humidity soared to broiling heights. To combat malaria, cholera, yellow fever, and disease brought on by inadequate food preservation or absence of sanitation, doctors relied on quinine, calomel, bleeding, and assorted patent medicines. Newcomers often fell victim to malaria, especially in the summers.

Travel proved difficult and sometimes impossible during or after heavy rains that turned normally dusty roads into brown squash and sent streams rampaging across the countryside. Partly because of these troubles, mail service in the early times (even after the days of the Republic of Texas, 1836–46) was slow to emerge from a haphazard level. On the Gulf, storms sank many vessels, which included a number of unseaworthy craft hurriedly put into service to take advantage of the rush to Texas, drowning immigrants and destroying freight intended for the new Texas market.

Such weather has figured in *any* history, of course, but Texas's past seems to have been influenced even more because of sudden changes and extremes. Attesting to this are letters and other documents in various archives, old and new books of history, and even roadside historical markers.

A Matagorda County marker tells a Civil War story with an impact of tragedy that has been largely erased by time. When the event occurred, though, many Texas residents were surely saddened for the rest of their lives.

On December 30, 1863, a group of Confederates led by Captain E. S. Rugeley (an old family name still widely known in the area) set out in "cottonclads"—gun-carrying vessels bulwarked with cotton bales—to drive off some Union soldiers posted at a beach position on the bay. The Confederates later embarked in skiffs and were attempting to land when a sudden norther lashed the bay and swamped their boats before help could reach them. Twenty-two men, including Captain Rugeley's seventeen-year-old brother, drowned or froze.

Weather dangers involved later in cattle-raising have been thoroughly recorded by participants and historians. But the stampeding effect of lightning bolts and booming thunder on trail-driven cattle can be visualized without being read. So can the results of severe blizzards in the cattle country, especially before fences appeared. Stock without shelter would turn backsides to the wind and drift with it. If the storm lasted long enough they would become icicled ghosts, stiff and gaunt, until they froze to death on the open prairie. After the arrival of fences, drifting cattle would pile up against those barriers and die together. Even wisely innovative men found themselves unable to cope with that kind of weather. Nineteenth-century buffalo hunters surprised by the sudden arrival of a deadly norther would sometimes cut open the carcass of a freshly killed animal and climb inside, as has been related in countless stories. But a few stories tell of still another surprise that followed: carcasses would freeze and hold the hunter in a frigid, morbid grasp.

Modern developments have eliminated exposure to some of these weather hazards, but in my own time—mostly but not solely in Texas—I have endured dust pneumonia, sickness because of spoiled food, near-heatstroke, skin cancer, two hurricanes, three near-misses by tornadoes, one painfall fall on ice, and still other injuries and illnesses and close calls I shall refrain from boasting about here.

It's enough to make me certain that the good old days couldn't have been so good, and to make me glad my experience with some of the early recorded Texas weather has been only through reading about it.

3. DEEP IN THE HEART

Texans bring to their varied climate a weather appreciation based on human history from its earliest. With the exception of some inevitable deviates, residents of this state usually find that the moderate days of spring and autumn are mostly to their liking. Their preference can probably be traced back to the first appearance on earth of humankind, which most scientists, creationist believers, and plain people instilled with logical reasoning seem to agree took place, in whatever fashion, in an area with annual mean temperatures that ranged between 70 and 80 degrees Fahrenheit.

Let the temperature rise ten degrees to 90 or more, and some Texans, especially well-fed ones, begin complaining of the heat. Let the temperature drop ten degrees, to below 60, and people begin complaining of the chill.

Other characteristics exist to indicate that humanity originated in a warm zone. Among these is our widely recognized loss of alertness and efficiency shortly after noon—around two o'clock, in fact. I had always thought a heavy lunch could have something to do with that, and I still think it might, but some research has shown the letdown to be natural and widespread and a hangover from humankind's earliest time in the original warm-weather habitat.

During the hottest part of those days of antiquity—an hour or so after high noon—activity surely ceased, as a climax to a late-morning peak of work. Another, but lesser, peak came in late afternoon. Today, people continue to show these tendencies, even when far removed from the tropics.

Scandinavian factory workers have been found to make more job mistakes in early afternoon, say anthropologists Maria J. Thompson and David W. Harsha, writing for the Field Newspaper Syndicate (February 21, 1984). German schoolchildren stumble

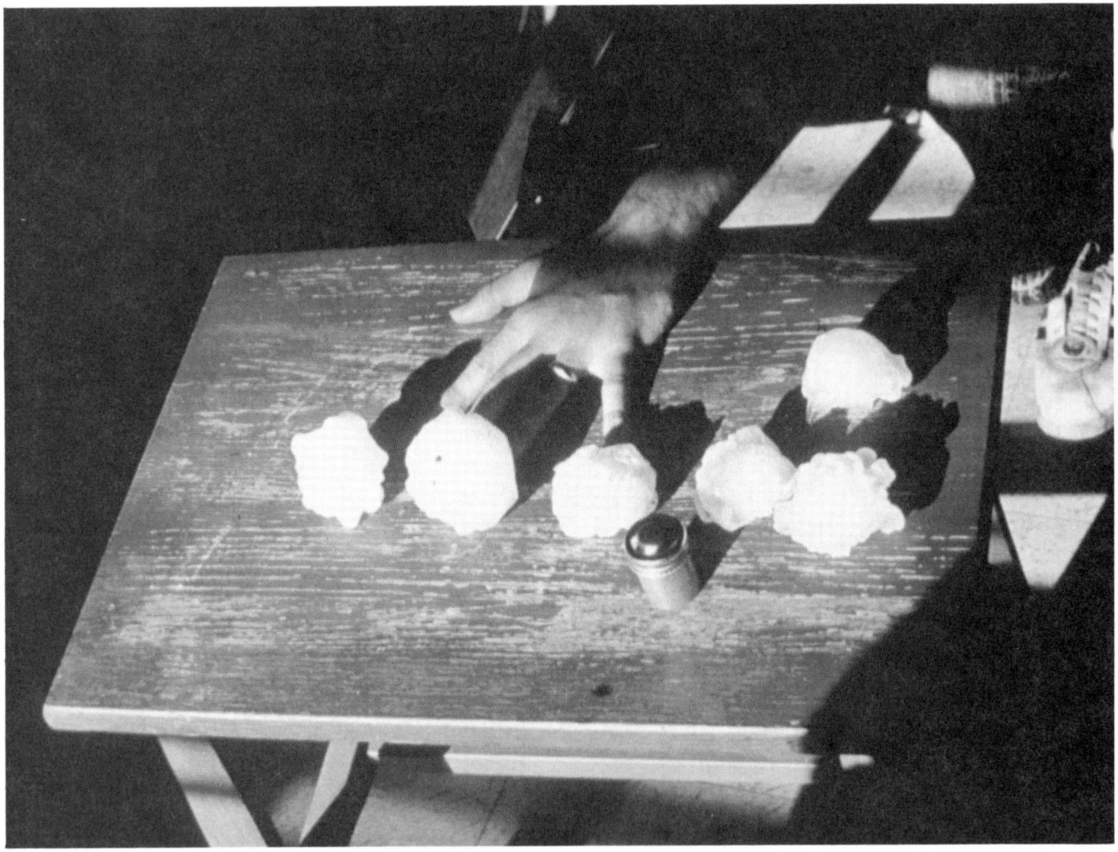

Hailstones 3 inches in diameter occasionally fall in Texas.

over postnoon arithmetic exercises. Rhythms from prehistoric days, not eating habits, are responsible, Thompson and Harsha claim, pointing to findings by circadian researcher M. J. F. Blake for proof. Blake has tested groups of workers for efficiency after eating at different lunch hours: 10 a.m., 12 noon, and 2 p.m. All three groups of twelve each still experienced an efficiency drop about the same midafternoon hour. Much other research manifesting a warm-climate origin for humanity exists. And it is apparent that most men and women are today most comfortable in a warm clime. This is fortunate in many ways for Texans, since their state is situated in what has become known as the Sunbelt, Amarillo winters notwithstanding. But it also brings to mind an old theory not very favorable to Texans: that people living in colder climates possess superiority of energy and leadership.

From the ranks of Texans themselves have appeared several able defenders. One was W. R. Woolrich of Austin, a man with whom I talked years ago while I was working toward a degree at the University of Texas. Woolrich was dean of the College of Engineering when he researched the very subject just mentioned, for an engineering study.[1] He remarked, "There is . . . no question but what cold climate conditions do create within man more urge to move fast, which in turn causes him to eat more high energy food. . . . [But] there is much evidence that less rigorous weather . . . is conducive to fully as effective creative action and thinking. . . . There need be no fear but what the people of the semi-arid areas of the Southwest can maintain their high energy position with the inhabitants of other parts of [the] United States or with northern Europe." Woolrich cited some historical evidence in defending warm-weather energies and abilities against the scoffs of go-getters from cooler climes. Did anyone, he asked, ever doubt the energy and ability of Apaches or Comanches—those sons and daughters of the Southwest who functioned with natural audacity and imagination without the benefit of air-conditioning? Nineteenth-century doubters who spent much time in the neighborhood of those Indians probably didn't live as long or as successfully as those wily people did.

Woolrich examined the data that he and others before him had accumulated and mentioned some other significant invalidations in that cold-versus-hot argument. Humankind, after its origin in a warm zone, migrated to colder climates only after the discovery of fire and the development of warm shelter and clothing. Providing motivation for this move were human curiosity and a desire to escape certain tropical dangers like diseases and vermin.

But moving into colder regions had disadvantages. It required that much time be spent gathering fuel, making better clothing and shelters, and adjusting to shorter crop-growing seasons. Still, the benefits of the move did indeed outweigh disadvantages—centuries ago, for sure. Those migrating people, who became known as Scandinavians, English, French, Germans, and so on, left behind them tropical handicaps, and they gained in other ways, too. They abandoned their nomadic tendencies and became a home-loving people devoted to family life, education, vocations. Their advancements gave them a superiority complex, and they came to look down on people living in warmer regions.

Today's world, however, is different. In

many warmer areas, especially in the southern United States, certain diseases that once resulted in human torpidity have been largely controlled: hookworm, malaria, yellow fever, typhus, dysentery. Other assistance has come from modern sanitary engineering and food preservation.

"Today millions of light-skinned people are living in the Southwest and Gulf Coast areas under semitropical conditions at relatively high energy levels," Woolrich wrote years ago—in 1952, long before the Sunbelt boom of the 1980s. (Thirty-two years later, data collected by the National Oceanic and Atmospheric Administration would corroborate this. In Sunbelt cities, NOAA found, the employee absentee rate falls the higher the temperature climbs, but in northern states the absentee rate increases with high temperatures.) "In fact," Woolrich continued, "a review of the last census reveals that the trek that started northward centuries ago has now reversed itself and more and more people are moving back toward the equator."

If you are wondering about physiological hazards involved in this, don't. People the world over have the same blood, at least in regard to enduring weather extremes. It can be differentiated by types, of course, but not by "thick" or "thin." Even Eskimos visiting the southern United States have been found to adjust to the new climate in a few weeks. It is simply a matter of getting acclimated—it's psychological, not physiological—and that's true, too, for a southerner moving north.

Many studies have substantiated this, but anybody can see for himself or herself. When the next frigid weather hits Texas, with the temperature dropping down to 20, 15, or 10 degrees for any considerable time, notice how comfortable a reading of 40 or 50 degrees will seem afterward.

Football coach Bud Grant, although no Texan, once gave further substantiation of the psychology involved in cold weather. While coach of the Minnesota Vikings, before the team moved into a protected stadium, Grant would tell his players that they could train themselves to ignore winter cold, and by doing so they could win games against opponents of weaker steel. He would use as one example a story about building early-warning radar stations in the Arctic, when American construction workers, actually overdressed against the cold, could not man outdoor machinery for long. Some imaginative construction boss brought in Eskimos, taught them the operation, and joyfully watched while those people worked long hours in continuing good health and cheer. After that, research scientists arrived to determine what made the difference. They examined body fat, blood types, and other things, and arrived at a conclusion: the Eskimos simply accepted the fact that it was going to be cold. Thus expecting it, they could work in it. The Americans refused to face that reality, fought it, and couldn't function.[2]

An Arctic explorer I knew, the late Vilhjalmur Stefansson, had learned this secret long before the wise construction boss. In a book, *The Friendly Arctic,* published decades ago, he explained how people from warmer climes could exist comfortably enough in the Far North by adapting to the Eskimo way of life and not trying to fight the cold. None of this means that anybody can ignore cold weather in safety. Infants six months of age or younger and people more than seventy years old are especially susceptible to chilly temperatures because their body mechanisms allow greater heat loss. And an extremely cold reading is not required for exposure to the dangers of hypothermia, or subnormal body temperature. Readings of 30 degrees or even higher, which can be recorded anywhere in Texas, often bring it on, because people aren't as careful about dressing warmly.

But all this chill is far from humankind's climate of origin and from the equator toward which so many Americans are moving in the late twentieth century. That equator is still far from Texas. The state's southernmost tip lies some two thousand miles north of it. Nowhere in the state, in fact, does the sun appear directly overhead at any time of the year, despite what you might think on a torrid day in August. The point nearest Texas where that overhead-sun phenomenon occurs is along the Tropic of Cancer, which is about 140 nautical miles south of Cameron County, at the southernmost tip of the state. But the sun approaches an overhead position sufficiently to build up plenty of steam for a Texas summer, beginning around June 21—the time of summer solstice. The season will assuredly be long and hot, no matter when it begins.

There is another certainty about the June solstice. It will be the longest day of the year in the Northern Hemisphere (and the shortest, of course, in the bottom half of the world, where winter will set in). Although the sun's angle will decline after that, and Texas days will begin shortening gradually, the heat will become greater for a month or two and possibly longer. Heat lag does it, the same as usually makes the warmest time of day two hours or so following solar noon. After the earth's surface has been absorbing and storing heat all morning (or all season) in a given area, it begins to give heat after solar noon (or summer solstice), but not so that it is noticeable at first to human sufferers.

In July and August, with temperatures continuing to climb throughout Texas, some inhabitants might find themselves wishing that a greater distance separated them from the equator. July and August are the only two months during which freezing temperatures have never been recorded by any weather observer in Texas.

The first howling norther of autumn or winter will, of course, reverse this dislike of hot weather and will bring back to many Texans, also including this one, a longing for those readings in the 70s or even the 80s of ancient preference.

Notes

1. "Recent Investigations of Human Energy and Related Design Factors for Hot Humid and Hot Arid Climates," *American Artisan* (July 1952).

2. Sports columnist David Casstevens wrote about this in the *Dallas Morning News,* January 30, 1984.

4. FORECAST: GREAT VARIATIONS

Many Texas days fall into the delightful range of 70 to 80 degrees, but this fact is often forgotten during the raging of weather extremes: a winter ice storm, extended summer heat and drought, a hurricane, hailstorm, dust storm, or tornado.

And Texas *does* have plentiful extremes. The weather can be different enough from day to day and from section to section to confuse and confound residents, who by now should know better. They can blame three peculiarities. Most obvious is the state's immensity. Its borders encompass three types of climate: maritime, along the Gulf Coast, characterized by mild winters and humid summers; mountain, in the Trans-Pecos area of far West Texas, with cooler temperatures generally than in the rest of the state; and, elsewhere throughout Texas, continental, characterized by sudden temperature changes and extreme readings.

Another reason is the variation in elevation, which ranges from sea level along the Gulf Coast to a sizable three- to four-thousand-foot plateau above the Cap Rock in the Panhandle region to that famed highest point in the mountainous Trans-Pecos area: 8,751 feet on Guadalupe Peak.

A third reason also explains much of the state's violent weather. The location between the towering Rocky Mountains and the low Gulf Coast has put Texas in an area where nature can rage. Cold air coming from the direction of the Rockies, warm, moist air emanating from the Gulf to the southeast, and high-moving warm, drier air flowing in from Mexico can converge in Texas. These ingredients, which even amateur observers know are as dangerous as some combinations in chemistry, can result in violence like severe thunderstorms and tornadoes.

Only in comparatively recent years have

"IF YOU DON'T LIKE THE WEATHER..."

A weather satellite view of Hurricane Alicia. August 17, 1983.

meteorologists realized the effects, too, of some other atmospheric phenomena: jet streams, personally encountered by American bomber pilots in the Pacific in World War II,[1] and Pacific moisture, which can flow into Texas across Mexico and become a weather factor.

The day-by-day and section-by-section differences have made most Texans much more concerned with their daily weather than with their climate, which can be defined as regional weather over a long period. A few examples will show why any averaging of figures could only prove misleading. Records of the National Weather Service (formerly the Weather Bureau) give Texas a 143-degree range between highest and lowest readings—and both were recorded within two hundred miles of each other, attesting to the temperature extremes possible in a continental climate.

At Seymour (southwest of Wichita Falls) the thermometer hit 120 degrees on August 12, 1936, long before the days of universal air-conditioning, and the actual temperature might have been even higher. Some years later an employee of the United States Weather Bureau inspected an instrument housing at Seymour and found it situated in the shade of a tree. He ordered it moved into the middle of a sunburned pea patch, but never again did the thermometer hit 120—or, at least, it hasn't *yet*.

That Seymour reading represents the highest accepted observation. Only two hundred miles away to the northwest, in Swisher County, lies Tulia, which reported the lowest official temperature recorded in Texas—minus 23 degrees on February 12, 1899. One other station forty miles beyond Tulia also reported a low of minus 23: Seminole, southwest of Seymour in Gaines County, on February 8, 1933.[2]

Another remarkable fact regarding the state's record high and low temperatures is that the same observer missed by only 9 degrees making both readings. Mrs. Claude Daugherty of Seymour, the woman who had kept the instrument box in the shade of a tree, had begun what would become a forty-five-year career as a volunteer reporter of meteorological data in 1925. Eleven years later, on that August day in 1936, she telegraphed the high reading, 120, to Weather Bureau offices in Fort Worth and Houston. Twelve years after that she sent a low reading from the same instrument box: minus 14 degrees on January 4, 1947.

Great variations in annual precipitation are recorded, too, but over much longer distances. Rainfall measuring 109.4 inches deluged Clarksville, in Red River County, in 1873; yet at several sites in the Trans-Pecos area amounts less than two inches annually have been recorded.

But the Trans-Pecos itself can have contrasting extremes. On June 13, 1984, Jeff Davis County residents were suffering through an extended dry spell. In the next six days twelve inches of rain fell, washing out bridges into Fort Davis. The county judge remarked, "We went from a drought to a flood in a matter of days."

Texas is one of the few states capable of having floods and droughts simultaneously. The year 1886 brought one of the worst droughts ever to western sections of Texas, encroaching even into Central Texas. Farm animals died by the thousands, and crops failed. But that same year also brought a roaring, wet hurricane that demolished the once-important coastal town of Indianola, and another hurricane brought Brownsville a thirty-inch rainfall for September.

"IF YOU DON'T LIKE THE WEATHER..."

In 1983, even before additional storm clouds had dumped more water on a soggy Houston, that city was said (on the "CBS Morning News" of August 18) to have received 220 percent of normal rainfall, while Wink, in West Texas, had recorded only 18 percent of normal. In one twenty-four-hour period during Tropical Storm Claudette in 1979, 43 inches of rain fell on a location near Alvin, in Brazoria County. During a September 1921 storm, 36.4 inches fell in eight consecutive hours at Thrall, in Williamson County, setting a national record.

Galveston Bay has frozen—on February 13, 1899. The Lower Rio Grande Valley has been in the extended grip of freezing weather several times—once for a one-week period (January 27–February 3, 1951): a total of ninety-three subfreezing hours, sixty-five of them consecutive, which killed three-fourths of the citrus trees. There have also been losses in recent years. In 1895 twenty inches of snow fell on the Gulf Coast at Houston, Orange, Stafford, and elsewhere. Six inches whitened Brownsville during that same period.

Still, a few persons have had the audacity to suggest that Texas weather really isn't singular. One is my friend Leon Hale, the widely read Houston newspaper columnist and novelist. In 1964 he wrote a column for the *Houston Post* that did indeed have some persuasive points. "Maybe the really singular thing about Texas weather," he declared, "is that even the natives have not learned to accept its normal activity." Hale said that people in other states probably have a habit of enduring similar sudden weather changes, but with less talk and boasting about them. He envisioned an Eskimo leaving his igloo on a cloudless night to toss his dogs a hunk of blubber before retiring. The next morning the man emerged to find igloos of his village nearby covered by snow. Hale said he doubted that the Eskimo would rush back inside and devote half an hour to telling his family about the occurrence, so common in the Arctic.

Maybe not. But if a wild wind had blown a fifty-degree temperature drop in three hours, especially from a starting point of 60 or 70 degrees, as some northers have done in Texas, the same Eskimo might have chatted about that—though maybe not in an igloo, which could have melted before the chill arrived. And if another kind of howler had sent dust billowing over a vast expanse of Arctic Ocean, as southwestern storms have done to the Gulf of Mexico, the Eskimo surely would have talked about the phenomenon of a dust storm at sea. This is not really meant to argue with Hale or anyone else who desires to retain his or her opinions about Texas weather. Other regions certainly can have sudden weather changes. Mark Twain is reported to have said, "If you don't like the weather in New England, just wait a few minutes."[3] Will Rogers has been quoted as having said the same thing about Oklahoma weather, and (as we all know) that's said every day, somewhere in the state, about Texas weather. Twain also said once he counted thirty-six different kinds of weather in New England in twenty-four hours.

All North America certainly does have great temperature variations and sudden weather changes. Yearly temperatures in New England can range from 110 degrees in the summer to 55 degrees below zero in the winter; yet it has been my own experience that Maine does indeed have only two seasons—winter and July—and during a visit one July 4 morning we had a wood fire

FORECAST: GREAT VARIATIONS

A mammatus cloud, under the anvil side of a thunderhead. This is a sign of turbulent weather.

burning in a stove.

Texas, because of its immense size (a fact that accounts for many other aspects of a Texan's outlook), can have some peculiarities of weather that are not noticeable in such numbers elsewhere. And Texas, also because of its size, can have some odd simultaneities, like those droughts in one area and floods in another. On the morning of December 16, 1983, I heard a nationally televised weathercast that reported Amarillo's temperature as 9 degrees at the same time that Brownsville was recording 71.

Like Leon Hale, Mike Kingston is a newspaper columnist who has attacked what he calls a stereotype of Texas weather: the dryness and dustiness of it, as usually seen in old westerns. Some parts of Texas get more than fifty inches of rain every year, he wrote in a 1984 issue of the *Dallas Morning News,* and Texas has forests, lakes, and coastline not often featured in films.

Quite true. But there is a great expanse of the state from, say, Fort Worth to El Paso and from Amarillo to Laredo and even beyond that can look very much like the scenery in those old movies. Such a panorama lay before my eyes almost every day until I was twelve. Some tourists in Texas see nothing else. Some *Texans* see nothing else.

Such cool professionals as meteorologists with the National Weather Service have remarked on the uniqueness and the hazards of Texas weather. Robert Ford of the Associated Press quoted one of them in a story dated February 8, 1970: "There are few places in the world where the weather is more restless, where air mass changes and local and regional-scale weather disturbances are more numerous than in Texas."

Eighteen years before that, Ford's wire service had given its own estimation of Texas weather, in newspapers of April 9, 1952:

"This is Texas for you:

"Tuesday it was 98 at Childress, the hottest in the nation.

"Wednesday morning it was below freezing at Dalhart and snowing.

"At mid-morning the Dallas U.S. Weather Bureau warned Texans to watch out for tornadoes.

"A half hour later several parts of the state were rocked by earth tremors."

So it's no Garden of Eden, as my farming grandfather learned long ago, but considering the heat, blizzards, volcanic eruptions, and earthquakes that strike random areas elsewhere, it's not a bad place to live, after all.

Notes

1. The Japanese used these jet streams to send thousands of bomb-carrying balloons to the United States in World War II. Although voluntary press censorship kept it quiet then, some 280 bombs hit ground—one as far inland as Iowa. Six persons died in Bly, Oregon.

2. Other high and low readings exist, but state climatologist John F. Griffiths and an assistant, Greg Ainsworth, wrote that one is surely in error and the other unofficial: 123 degrees at Fort Ringgold, Starr County, in 1879 (wrong), and minus 30 degrees in Ochiltree County in 1899 (unofficial) (*One Hundred Years of Texas Weather, 1880–1979). Higher temperatures have been reported, too, with hot winds that sometimes destroy crops in a matter of hours; but these are unofficial.*

3. *Folklore of American Weather,* by Eric Sloane (New York: Hawthorn, 1963). Twain also said, "Everyone talks about the weather, but nobody does anything about it"; but he picked *that* up from a man named C. D. Warner, who might have heard it somewhere else, himself.

5. SKY WATCHERS

Fact is sometimes hard to separate from fiction in talking about Texas weather, and sometimes the two can even merge. There's the *true* story about the mercury that, one winter, dropped three feet in a few seconds when a norther blustered in. Gusty winds knocked the thermometer off an outside wall where it had been hanging.

But fact and fiction aren't likely to merge in some other tall, hackneyed tales of Texas weather, such as the one about another fast-moving norther that dropped temperatures so quickly that it froze ducks and geese in a lake where they sat, until hunters shot at them and sent ducks, geese, *and* frozen lake flying into the heavens.

Nor in some presumably original one-liners solicited by the *Austin American-Statesman* from subscribers to commemorate the cold, cold December of 1983 (and published January 8, 1984): "It's so cold the birds are flying north this winter" (Lisa Blech); "It's so cold that when I opened the garage door this morning, the car pulled it back down" (David Blaha, Jr.); and "It's so cold I didn't know I had a flat tire until it thawed" (David D. Haun, Round Rock).

Nor in the not-so-old tall tale contributed by Pat Folmar Robinson of Austin, who told about an old Texan marooned in his residence by one of those infamous flash floods that can strike almost anywhere in the Lone Star State. Firemen quickly came by in a boat, evacuating residents, but the man refused to leave. "I'm staying," he said. "The Lord will take care of me."

But the water rose further, sending the man to his second floor. The same boat returned with another offer of help. "You've got to get out," a fireman shouted.

"No," replied the man, "the Lord said He'd take care of me."

Still the water rose, and the old man climbed onto the roof. Crewmen in a rescue helicopter spotted him and one shouted down through a bullhorn, "We'll throw down a rope ladder."

"No!" the man yelled back. "The Lord has a plan for me."

After the house had collapsed and the old man had drowned, he demanded of St. Peter an audience with the Lord. "How come you let me drown, Lord?" he asked.

"Well, I sent you two boats and a helicopter," the Lord answered. "What more would you have wanted me to do?"

Most of the tall tales have been left out of this book. Such was the hope and intent, anyway. The stories narrated here are true, or meant to be true, unless labeled otherwise. But they can border on the incredible. Texas weather and absolute confusion can sometimes be closely related.

"When you talk about normal weather in Texas, it's a misnomer—almost," Bob Riggio, an official with the Texas Department of Water Resources, has said. "Abnormal weather is the norm."

The year in which he said that went out with a series of tornadoes in December and at least one hurricane threat that same month. Tornado season usually peaks in May or June and hurricane season in September. This is the state where a lake near San Angelo blazed up during a drought some years ago when vegetation on the baked, waterless bottom burned, forcing local firemen to rush forth and put out the "fire on the lake," and where, years later, a rattlesnake somehow appeared in a commode in the same drought-plagued city, terrorizing the homeowners and requiring the help of an expert to get rid of it.

Weather forecasting for Texas has become a tricky thing, but this has been true elsewhere of weather forecasting, dating back countless centuries into humankind's earliest history.[1]

For thousands of years people have been putting their weather observations into proverbs, and I can understand why. Living with an absolutely unknown and unpredictable tomorrow would have become unendurable to the first men and women, especially when weather governed their existence to the extent that it did then. A bit of silliness described its grip:

> "What is it molds the life of man?
> "The weather.
> "What makes some black and others tan?
> "The weather.
> "What makes the Zulu live in trees,
> "And Congo natives dress in leaves,
> "While others go in furs and freeze?
> The weather."

Eskimos don't "go in furs and freeze." They might dress in furs, but they live in harmony with nature (like other natives of any time and place), and thus they usually live in reasonable comfort, as explorer Vilhjalmur Stefansson wrote in *The Friendly Arctic*. Still, the verse makes its point: weather has pretty much owned mankind at least into the twentieth century, and it possesses us today in ways we might not realize.

Several scientists, including Dr. Helmut E. Landsberg of the University of Maryland (and former director of climatology with the United States Weather Bureau), recently discussed these ways. Among arthritic patients, rising humidity and a falling barometer—which affects fluid-filled cavities in body joints—will cause greater pain. People suffering from respiratory ailments will find breathing more difficult during a period of air stagnation, when lack of

Streak lightning. Lightning bolts from one-half to one inch in diameter can contain from 100 million to 1 billion volts of electricity, and the peak temperature can rise to 50,000 degrees Fahrenheit—four times as hot as the sun.

wind can result in atmospheric pollution. Individuals subject to unusual mood changes might find dark, overcast days exceptionally depressing and the long nights common to parts of the globe impossible to endure. For some persons a period of hot, humid weather can ignite violence or suicide. Sunshiny days encourage migraine headaches. Temperature fluctuations—not extended low temperatures—bring on the common cold, probably because of an ensuing strain on the heat-regulating system of the human body that results in chilling and in weakening resistance to germs. Warm weather causes a drop in blood pressures of people with hypertension; cold results in an increase.

These are some of the physiological reactions to weather. We all know the more obvious ways in which weather holds sway over people—for instance, in the winter, when shoveling snow can result in heart attacks and hazardous driving conditions in injuries and fatalities. The National Weather Service has kept records of such deaths and has found that three thousand Americans died in winter storms from 1936 to 1969. Fewer than four hundred deaths, however, were direct results of exposure. The rest came in automobile accidents (one-third of the total) and in other ways.

"What is it molds the life of man? The weather." It has done so, and it still does. We can all appreciate the reason for anxieties among the earliest human beings about what tomorrow might bring. Fishermen a thousand years removed from today's tuna and shrimp fleets and their access to scientific forecasts needed to know whether they would be sailing on a clear, calm Monday into a stormy Tuesday or Wednesday. Centuries ago, farmers, too, benefited in several obvious ways by having advance notice about the arrival of rain. Any of our remotest ancestors gained by being warned of impending weather changes. Thus the wisest and probably longest-lived of them became sagacious observers, as did, in time, my farming grandfather.

So their rules for foretelling various changes eventually went into proverbs. Some early examples have been found on Babylonian clay tablets dating back to 4000 b.c. A number of proverbs are applicable even today, anywhere in the world. They were originated by skywatchers who had successfully determined the actual order of weather phenomena, although they couldn't even guess at causes. Other weather sayings have proved valid only for the particular localities where they were formulated: areas in valleys or near mountains or water that influenced weather conditions in the immediate vicinity and rendered proverbs from those places misleading or meaningless for the rest of the world. Finally, a number of maxims based on superstition, fancy, or ignorance (like the story of the groundhog and its shadow) are known as totally false, although coincidences might sometimes make them seem valid.

Texas has had its share of valid and invalid proverbs, and sometimes it becomes quite a task to sort out which is which.

In 1883 what is now the technically sophisticated National Weather Service, but then was part of the Signal Service of the United States Army, collected and published *Weather Proverbs,* without any great effort to confirm validities and with some scoffing from scientists. But the officer who prepared it (H. H. C. Dunwoody, First Lieutenant, 4th Artillery, the assistant signal officer under Brigadier General W. B. Hazen) remarked in an introduction,

> Many of these sayings express in a crude form, the meteorological conditions likely to follow, and have resulted, from the close observation on the part of those whose interests compelled them to be on the alert, in the study of all signs which might enable them to determine approaching weather changes. . . .
>
> The ablest meteorologists of to-day, aided by the most perfect meteorological instruments and the results of years of accurate instrument observations, are still unable to give reliable forecasts of the weather for a longer period than two or three days, and frequently not longer than twenty-four hours. It is possible that a more accurate observation of plants or the condition and action of animals might lead to some valuable suggestion in this important field of investigation.

That last paragraph is not so antiquated as its date of 1883 might make it seem. What Texan can doubt that the flying of geese southward presages colder weather? And, as for long-range forecasting, even today's meteorologists admit that it has accuracy limits, and that three, four, or five days, and often fewer, are the maximum range for authenticity (although better computers might change that before long). Sometimes I have wondered if television weathercasters have even looked outside while studying computer and radar information before giving a forecast. (Obviously they don't *have* to.)

Many of those old proverbs, but certainly not all, had validity, having been based, for example, on nature's computer, when "aqueous vapor" acted on animal and vegetable organization. "The increase of aqueous vapor is indicated by the expansion or contraction of various substances, such as wood, whalebone, cat-gut, sponge, and hair, which, when colder than the air, condenses the moisture upon them, and this being absorbed increases the temperature, thus causing expansion or contraction."

Here, then, are a few old proverbs from the 1883 Signal Service book that, among many others, could apply to Texas with varying degrees of authenticity.

> A coming storm your shooting corns presage,
> And aches will throb, your hollow tooth will rage.
>
> When gnats bite keenly, and when flies keep near the ground,
> We look for wind and rain.
>
> When cats are snoring, foul weather follows.
>
> When cattle hasten to timber, expect a "norther."
>
> The hair of a horse appears rough just before rain.
>
> Horses and mules very lively without apparent cause indicate rain.
>
> Animals making unusual noise indicate change of weather.
>
> When sea-gulls fly to land, a storm is at hand.
>
> Fish swim upstream, and catfish jump out of water before rain.
>
> Bees will not swarm before a near storm.
>
> In Texas and the southwest when the wind shifts during a drought, expect rain.
>
> Gentle or brisk east wind precedes rain. Southwest or west wind indicates the approach of clear, dry weather.

And so on.

In 1923 J. Frank Dobie wrote a piece for

the Texas Folklore Society entitled "Weather Wisdom of the Texas-Mexican Border." He began by quoting Solomon—a verse in the Bible from Ecclesiastes 11:4 that is "finely said," Dobie remarked, "and no doubt prefigures a truth; but the wisdom of Solomon was not weather wisdom." The quotation: "He that observeth the wind, shall not sow; and he that regardeth the clouds, shall not reap." Obviously that is not weather wisdom, but as Dobie himself certainly knew by reading it in context, it was not meant to be that. Dobie was, however, on the right weather track. He intended to say, of course, that successful farmers must observe the weather—or, in my grandfather's case—even often-unsuccessful farmers who hope for improvement.

"Folk who herd and harvest have always been weather wise," Dobie said. "Thus, the Mexican *vaqueros* and *pastores* who tend their pastures and herds on the border lands of Texas have come to possess a weather lore of signs."

Many signs centered around rattlesnakes —creatures that, according to popular notion, avoid getting their rattles wet. One weather belief had it that their emergence from dens, or wherever the serpents might consider home, heralded rain, especially if the snakes then headed for high ground. In that case, Dobie wrote, "an old settler with an optimistic turn of mind might go out and buy a herd of cattle." Dobie himself indicated that he believed the veracity of this folktale, and he *knew* rattlesnakes. Although they swim in water as naturally as does the moccasin, he said (and he told of once watching two rattlesnakes slithering around playfully in a water hole "apparently for no other reason than for fun"), they do, he added, keep their rattles pointed upward, out of the water. Even as Dobie wrote his article, however, he read a story in the *San Antonio Express* (datelined Bishop, in the Rio Grande Valley, on an April 15): "The country is alive with rattlesnakes. . . . Old-timers predict that they are the forerunner of another [drought]."

Whatever the actual significance of those crawling rattlesnakes, old-time western fiddlers knew the legend. They put rattles in their instruments as a charm against damp strings: while attached to living things the rattles, at least, had been kept dry.

The people of Mexican ancestry whom Dobie knew and wrote about in the article were great believers in weather signs and wonders. Sometimes they could believe with too much faith. When coyotes howl every morning after sunrise (but not after sunset or before sunrise), it will rain, they believed. When owls hoot in the daytime and burros bray after dusk, it will rain. When lambs display friskiness in the morning, it will rain.

But those Hispanics, like other western Texans, usually needed rain, and they were occasionally too optimistic, too easygoing, or too willing to please others with their weather forecasts. One night, Dobie said, he observed a milch calf who refused to suck.

"What is the matter with it?" he asked the old Mexican who was milking.

"*¿Quien sabe?*" the man replied. "It is probably going to rain."

The border people Dobie knew seemed to think that every animal in existence provided weather signs at one time or another. "Once as a boy," Dobie wrote, "I asked a Mexican what the unusual number of bull-bats that we saw flying about in late evening was a sign of, and I remember well my astonishment at his reply: that they were not a sign of anything."

Dobie never was a man to lose himself in blind faith, as anyone who knew him realized in a short time. He recognized many of the border weather signs as well founded, and even in 1923 he saw the logic for this. If ranch cattle and other animals could indeed sense a change in the weather, and this surely seemed so, it was largely because of changes in barometric pressure and other realities. He was, otherwise, no believer in signs, wonders, and things like the power of organized prayer or Indian dances to bring the rain.

"Texans have a saying," he concluded. "In dry weather all signs fail. . . . And the most prolific Mexican weather prophet often admits that it will rain only *'cuando Dios lo quiera'* (when God wills)."

Despite his cynicism regarding humankind's beliefs and claims, Dobie enjoyed talking with all kinds of people (often excepting professional educators, journalism teachers, and preachers), and he surely would have met a game warden from Colorado County, Tom Waddell, who visited Austin several times —once on September 9, 1953, when Bill Brammer[2] interviewed him for the *Austin American.*

Earlier that year, on August 13, 1953, Waddell had surprised Texans with a daring prediction that proved correct. During a prolonged drought, with temperatures consistently registering higher than 100 degrees, Waddell had forecast an end soon to such unpleasantness. If he hadn't been surprisingly accurate with some earlier forecasts, nobody would have paid any attention to him. But a reporter wrote a story about the prediction, and it went around to Texas newspapers.

The same Austin newspaper Brammer worked for had, in fact, headlined the story: "Relief from Heat on Way," beneath a small overline-headline: "If You Believe Ducks, Deer."

"Most folks didn't believe the ducks and deer," Brammer said later of the forecast. "Even the weatherman held to his same prediction—no relief in sight."

Waddell had remained firm. He had observed fawn beginning to lose their spots, then buck deer rubbing velvet off their antlers two and a half weeks earlier than usual. Waddell knew for himself that hot weather was nearing its end, even while other Texans (like this author) sweated and complained.

Within a week of the incredible forecast a norther blew in. It wasn't a strong one—it came in *August*—but it brought rain and a twenty-degree temperature drop. Never again that summer did thermometers hit 100—at least not in Waddell's general vicinity—and three weeks later a genuine norther moved in, bringing up to five inches of rain to parched areas of Texas.

Many observations and predictions like Waddell's are obviously very well founded, even if the exact reasons sometimes might not be known. In early days of Texas, for instance, when wolves were more numerous in areas now densely populated by humans, it was said that "wolves always howl more before a storm." If that was indeed true, and it seems to have been, it might have been because their ears hurt. And if their ears hurt, it was probably because of the previously discussed low pressure, which brings weather changes.

Low pressure can also cause birds to walk when they would ordinarily fly, because the air change requires greater effort

The tail end of a roll cloud, sometimes mistaken for a tornado funnel.

A typical tornado. Tornado season in Texas usually peaks in May or June, but tornadoes can occur at any time of year.

for flying. Insects also fly lower, swarm, and bite harder, and shortly before a storm strikes they might quit flying altogether.

Low pressure can even be smelled, some people contend, and my mother, a classical musician and a very sensitive woman in many other ways, too, declared this to be so. She maintained she could smell an approaching storm (although she probably didn't know why; possibly it was because low pressure allowed certain natural scents to spread out).

Many of us have similar beliefs and prejudices about the weather. On October 31, 1984, my friend Jacqueline Simon, a poet-novelist who lives in Houston, wrote an interesting reply to a query from me about weather experiences.

> I always rev up in the fall, when a little relief after the summer blows in. I read just the other day, in Tolstoy's daughter's biography of him, that he got his best work done in the spring, when relief after the harsh winter appeared. Another interesting fact about weather (I learned this doing some work on schizophrenics) is that a far disproportionate number of the severely mentally ill were born in the last, severest, months of winter—in our hemisphere, January and February produces the most psychotics; in the southern, July and August. No reason can be determined for sure, whether it's greater exposure to viruses, say, or what.
>
> Also I have learned, although I'd be hard put for the source for this, that the balmy spring months produce the most intelligent babies, as well as the healthiest. One reason for that might be that the infants have the long summer months to grow strong for the winter. (This might have been true in the past more than now.)
>
> I am very attached to this theory, having been born in April, but I think the time has now come to confess that I once examined one of those calendars that records the birthdates of history's most outstanding folks, with an eye to finding most of them clustered between April and May. The truth is that they are about as evenly divided as can be. . . .
>
> According to my research more people *die* in January than March, but no doubt that is because they are thrilled to death to get the Christmas shopping finished.

On my desk in front of me right now I have a stack of items concerning weather and climate psychology. Here is a small sampling, with names deleted in some instances.

A group of Dallas psychologists found in 1984 that more suicides occur in that city in the spring than any other time of year—because, said one, people don't kill themselves while in the depths of depression. They are immobilized at that time. Suppose the greatest depression arrived during the Christmas season, as is often claimed. By spring the depressed person can act on the determination to end his or her life.

But about the same time another psychologist, who resides out of state, said the well-publicized theory about Christmas leading to increased depression and suicide is not valid. His data indicated otherwise, he said. But he *did* happen to corroborate, in a way, some of the previous theory (about an interval between death thoughts and action) by remarking that December is the lowest month for suicide, April the highest.

Most babies in Houston, where the following story originated, and in many other United States cities, are born in August and September, a group of researchers discovered in 1984. Immediately a good reason for this became clear. Cold weather in November and December had kept more people at home, apparently snuggling in bed—but in my clipping files I find some disagreement. A Florida psychologist, while agreeing that more babies are conceived in chilly months, also contends that cold weather makes people more fertile as well as more sexually active.

Things get mixed up, too, in regard to the croaking of frogs. Forecasters of the folklorist sort say this foretells rain; scientists sneer and say it's a mating call.

Take your pick of these theories. Regarding the human puzzle, I am inclined toward the snuggling explanation. As for the frogs, let them make up their own minds.

There is another weather theory, a logical one, that hot weather leads to violence; yet workers and directors of shelters for abused women and children around Texas said in 1984 that family crises increase as the temperature falls. The inclement weather is a reason, of course, but the direct cause is stress—stress ensuing from being cooped up, from frozen water pipes and automobiles, from ruptured utilities, and so on. Texas Department of Human Services workers were quoted as saying that January and March are two of the most active months for complaints about child abuse.

Most requests for admissions to Texas mental hospitals also come in winter months, state workers said, and deaths from drinking alcohol peak then. However, crimes like burglaries and armed robberies drop, apparently because of the criminals' discomfort in the cold.

Still another clipping quotes a Texas authority on how weather affects human beings, and he has a lower regard, generally, for such influences. No hard evidence exists to support the belief that normal weather changes bring on cardiovascular and respiratory ailments and other illnesses, he asserted. Weather alone, for example, cannot be blamed for colds, said the man—Texas A&M University professor Dennis Driscoll.[3] Possibly your cold came from wearing insufficient clothing in cold weather, or maybe being cooped up in an office where someone spread the germs—an easy process in a tightly sealed building. Blaming the weather for illnesses gives people a needed scapegoat, he said, and allows them to absolve themselves of responsibility.

The classic example of blaming weather came during polio epidemics years ago. People blamed summer weather until scientists discovered the real reason for those epidemics: an organism that lived in water and infected summer swimmers. If there had been as many winter swimmers, Driscoll speculated, there might have been as much winter polio, too. Driscoll traced the "mystique" regarding the effect of weather on health back to Aristotle, the Greek philosopher who wrote *Meteoroliga,* widely used as a textbook until two hundred or so years ago.

Aristotle lived from 384 b.c. to 322 b.c. Go back in Greek history a hundred or so years before that and you'll come to Hippocrates, the Greek physician known as the father of medicine, who lived from about 460 b.c. to 377 b.c. and who also had something to say about weather and its effect on humankind, whether right or wrong. Hippocrates advised that an aspiring physician should study meteorology before studying medicine, because the changing seasons, hot and cold winds, and the stars can have varying effects on the human body and its health.

Dating back those multicenturies is the popular notion of being "under the weather" when in bad health. That connection between weather and health is a reminder of a remark made to my wife by a Waco surgeon who performs bypass operations. He used to "get mad" at people who told him they could foretell weather changes from their wired-together cardiac scars. Now, after having had bypass surgery himself, he claims that he can foretell the weather this way, too.

Psychological studies that show a link between weather and human actions also impress me as being logical. A Rice University professor, Craig Anderson, was quoted in the *Houston Chronicle,* April 12, 1984, as saying that violent crimes like homicides and rapes increase in Houston as heat increases, but he qualified this by remarking on other possible reasons. Rapes might increase, he said, because more women stay out late or decide to take evening walks, thus increasing the chances of harm.

In 1954 Dallas district attorney Henry Wade observed that violent crimes jump in number when the thermometer goes over 100 degrees and stays there for a while, although professional crimes (like safecracking) remain at the same level all year.

High winds blowing over an extended period can cause an increase in accidents and suicides as well as crime. Most history-minded Texans remember reading about another effect of wind: Walter Prescott Webb's description of the torment it caused women on the plains. I recall hearing him discuss this in some detail in a course on the Great Plains at the University of Texas. But some psychological studies have shown that those same winds can affect men, too, sometimes causing (in addition to accidents, suicides, and crime) sleeplessness, irritability, tension, and migraine headaches accompanied by nausea and vomiting. This, too, seems only logical.

But one of the most remarkable studies I have read is of a weather cause not so commonly observed: "Rain and the Urge to Urinate: Clinical Observations Concerning External Events, Surmounted Impulses and Repression," by Owen Renile in the *International Journal of Psycho-Analysis,* 1981—a London publication "founded by Ernest Jones under the direction of Sigmund Freud." The study took place in California, not Texas, but people are, presumably, much the same everywhere, whether or not inhabitants of either of those two states care to admit the mutuality of this.

In California, in the winter of 1977, rains began falling after three years of drought. At first came a general feeling of relief; but, as residents saw no sign of a letup, they began to feel discomfort, then worry. Flooding became a constant threat, driving a miserable chore (more so than usual), wet clothes commonplace. Anxiety caused people to sleep only lightly at night, and water imagery became prominent in dreams. Patients told the psychiatrist who wrote this article of their worries about a possible loss of control over their impulse to urinate. Several mentioned repeated feelings of a strong desire to do so, but when they went to the bathroom very little ensued. Most of the action continued to take place outdoors.

Notes

1. See *The Tornado,* by John Edward Weems (New York: Doubleday, 1977) for narration in detail.
2. Another Texas writer of my long-time acquaintaince, now deceased; author of *The Gay Place.*
3. United Press International story, May 23, 1982.

6. ISOBARS, CROWBARS AND BAD ENGLISH

The preceding weather studies come with no guarantee from me, although most of them are, obviously, scientifically founded. But neither do contemporary forecasts come with my warranty.

Today's weathermen have taken criticism aplenty, much of it unjustly. The National Weather Service does a good job, considering the lingering inexactness of the science of weather forecasting, and it has endured unfair criticism—extending back at least as far as 1900, to my personal knowledge, and no doubt before that—long before the days when it took its present organizational name.

In 1956, while researching the disastrous 1900 Galveston hurricane, I heard complaints from several surviving Galvestonians that the Weather Bureau gave no warning of danger. This was not true. Weatherman Isaac Cline tried to warn people along the beach to seek safety, but the poor communications of that day interfered with the dissemination of information. He was reduced to driving up and down the beach in a horse-drawn two-wheeled cart, shouting his warnings.

And at least one remarkable triumph of weather forecasting stands out in my memory, although there have been countless others, along with the usual number of failures. At 9:30 a.m., May 11, 1953, when specific forecasting of tornadoes had just begun, the New Orleans Weather Bureau office issued this forecast after consultation with the Washington, D.C., office: "There is a possibility of a few tornadoes in the [Texas] area bounded by San Angelo, Waco, Wichita Falls, and Big Spring."

Differences existed then among weathermen about the advisability of predicting tornadoes. The Weather Bureau had been able to issue warnings since 1942, but the earliest ones were based only on reports of individuals who had actually sighted funnels. Advance predictions were not attempted until 1952. Even then, some officials believed that forecasts naming specific areas of possible storms would only terrify the people concerned and do more harm than good. Another reason for caution was explained by Francis W. Reichelderfer, chief of the Weather Bureau in 1953. "The Bureau occasionally gave out information [earlier] that the weather map indicated tornadoes, but we felt that forecasts weren't accurate enough to publish as a definite feature until [1952]."

That May 11, 1953, tornadoes ripped through two of the cities, San Angelo and Waco, mentioned in the warning out of New Orleans. I have since discovered that several meteorologists participated in the preforecast discussion, including I. R. Tannehill, chief of the forecast and reports division. But the warning as released was signed simply, "[W. C.] Conner, Weather Bureau."

The Waco office of the Weather Bureau that day was in the charge of a member of the old school of weathermen, C. A. Anderson. He was kindly and thoughtful, as I was to see later when interviewing him, and he was well liked for his humanity. He did not believe in frightening people needlessly, and he sought to reassure Wacoans after other meteorologists had issued such a startling, precedent-making forecast. Later he would regret that attempt until he died.

The local *Tribune-Herald* quoted him in regard to the warning of possible tornadoes:

"In Waco, Weather Forecaster C. A. Anderson said there was no cause for alarm in Central Texas. Anderson said at the time [1:30 p.m. May 11] that no tornadoes had developed in the area, and that if any did, they would travel over only a small area. He said they would be west of Waco if they did strike.

"He predicted strong winds, scattered thunderstorms, and probably some hail in limited areas."

So the headline in Waco that preceded the dire weather forecast out of New Orleans proclaimed, "'NO CAUSE FOR ALARM' HERE." About 4:30 that afternoon, even while copies of the newspaper were still being sold, at least 114 Wacoans died when the tornado struck downtown areas.

Today's forecasts are not tempered by such kindliness, but they can still be flawed or downright incorrect, and they have even evoked at least one lawsuit (for $2 million, though not in Texas), when an error resulted in the deaths of three fishermen lost in a boat.

Numerous hazards exist even in this day of automation to make weather forecasting an inexact science. A weather system's speed of movement is difficult to determine, as is the amount of moisture held; yet accuracy (not only in recording current conditions, but also in estimating developments) is essential for making a correct forecast. And when the temperature hovers around the freezing point of 32 degrees, many different things can happen. Let the temperature drop a bit below that, and snow, sleet, or ice pellets might fall. If the temperature rises a little, rain will result. Often, of course, there is a mixture.

Many other difficulties present themselves. Today's computers are not sufficiently accurate, especially when incorrect information gathered by mere men and women goes into them. Despite all this, accuracy is said to have improved to a current

ISOBARS, CROWBARS AND BAD ENGLISH

J. H. Jarboe, San Antonio's "weather prophet" stands beside a nephoscope, a black mirror which reflects the clouds and measures their velocity and direction. January 12, 1926.

rate of around 85 percent or a bit higher (not including long-term forecasts.)[1]

"The meteorologist, like the physician, must be part scientist and part artist," Texas meteorologist Vance Moyer has been quoted as saying. "Unfortunately, you cannot train someone to be an artist."

So there is plenty of imperfection, and at the least this brings forth complaints, if not lawsuits. The chief of the National Weather Service, Richard Hallgren, understands why. "Public griping is justifiable because we don't do as well as they wish and need," Hallgren said in an Associated Press story January 9, 1985. Other meteorologists disagree, saying that the public has no right to expect perfect forecasts as long as many intimate mechanisms of weather are not known.

At the time Hallgren made his statement, the Weather Service was able to give advance warnings of only one out of three large tornadoes and one out of two flash floods. But he predicted near-perfection in the not-so-distant future in giving severe-weather warnings that could help save lives. Computer improvements would help facilitate this.

These days the Weather Service has no monopoly on forecasting, as it once had. Private services have developed; they base their predictions on data supplied by the National Weather Service to anyone who wants it. Some of those services forecast for distant clients. For Houston this has resulted in an estimated 390,000 radio listeners getting many of their forecasts from hired private meteorologists broadcasting from cities as far away as Boston (*Houston Post,* May 22, 1983). The out-of-state forecasters claim a great degree of accuracy, because all National Weather Service information recorded is available to them, but some authorities, such as Texas climatologist John Griffiths, point out that on-the-spot forecasts are better because unexpected changes can affect local weather quickly.

Others say the same. Fred Sackett of Austin, who grew up on a ranch near Coleman and later served as an aviator in World War II (and thus has a double interest in the weather in his background), wrote me:

> It seems to me that the quality of weather predictions began to slip when, sometime in the 1950s or 1960s the [Weather Bureau] began feeding data into a large computer. . . .
>
> I maintain that the variables of even a simple weather system are too complex and difficult of measurement to lend themselves to a mathematical analysis except of the most basic sort. Hence forecasting is as much an art as a science. Once U.S. Weather Bureau forecasters spent most of their careers in one locale, thus developing a feel for the nuances of vagaries of the area's weather. No more! Most of the NOAA people here in Austin are from out of state. They consistently miss, for example, temperature predictions, for they apparently can't accept the rapidity of our winter warming trends or the precipitous drop in temperature upon the passing of a "blue norther."

For the author of this book any discussion of forecasting methods and computer developments is a hazardous thing and probably should be dropped at this point, or maybe should have been dropped even earlier, for two reasons: information could easily be out of date before the book is published, and this author's expertise does not encompass computers. He knows more about those old skywatchers, being the grandson of one of them.

He does claim, however, some knowledge of television weathercasters, having seen and listened to them often enough, and he holds some of them in less esteem than the National Weather Service people, although there does seem to have been development in professionalism recently.

Fred Sackett also has some remarks on the subject. Originally he presented them in a letter to the editor of the *Austin American-Statesman:*

> As an old Army Air Corps pilot, and a lifelong weather nut, I wish to add my voice to James C. Fidler's lament (letter, Aug. 1) on the sad state of weather reporting on Austin television. . . .
>
> The pap which is now fed to us by incompetent exhibitionists is literally incredible. Not only do they not know an isobar from a crowbar, but they tend to butcher the English language. A few months ago one of them repeatedly asked (I quote), "How fast does snowflakes fall?" He then went on to give a precise rate as if all types of snow have the same terminal velocity! This same man recently "explained" high and low pressure areas by saying that the air rises in low pressure systems, thus pulling up and decreasing pressure, while it falls in high pressure systems, thus pushing down and increasing pressure! The simple fact is that pressure depends only on the weight of the atmosphere above the point of measurement. These simpletons rarely mention upper level winds or jet streams which have such profound effect on our surface weather, and they probably don't even know about the undulations of the circumpolar vortex which alters jet stream patterns.
>
> The real culprits in this sad saga, however, are the station managers who seem to prefer posturing and falsetto voices, or pretty faces, to scientific weather reporting [August 11, 1983].

My own preference is also for a serious television weathercaster, not one who is a flippant show-biz buffoon. After all, the weather is rarely a joke, usually not even to large-city residents with access to all their weather modifiers. Blizzards, floods, tornadoes, and hurricanes can still bring city business to a halt—and unless Texas does something effective about its frequent water problems, the time might even come when city-dwellers turn on water faucets and see drops of nothing come out. That's when many of them will learn to enjoy, as I have said I usually do, a slow, drenching rain. And that's when fewer of them will put up with the clowning antics of TV weathermen.

Some television weathercasters are, of course, trained meteorologists, and they are more likely to act professionally. Many other forecasters are simply station employees apparently hoping to host the "Tonight" show at some future date; in one case some years ago such a weathercaster jokingly threw darts at a board to get his forecast. I hope these are exceptions, but I haven't watched every weathercast, not even in Texas.

One broadcaster who has been impressive (and, of course, there are others) is Harold Taft of Fort Worth. Taft has a meteorology degree from the University of Chicago, where he was sent by the Army Air Corps during World War II. A retired colonel now, he briefed generals Dwight Eisenhower and Mark Clark and officers of the Strategic Air Command during his military career. Since 1949 he has reported on the weather for Fort Worth/Dallas radio and television. "I do the weather just like I did when I was back in the military," Taft has said. "If you're looking for glitter you've

"IF YOU DON'T LIKE THE WEATHER..."

Instrument used by J. H. Jarboe, San Antonio's "weather prophet," to record local weather conditions. January 12, 1926. Modern weather forecasting methods using radar and computers present a striking contrast to this type of instrument.

come to the wrong place." A "High Profile" feature in the *Dallas Morning News* of May 22, 1983, tells more about him.

Other weather forecasters who go about their television duties with some seriousness are those approved by the American Meteorological Society. To get the "seal of approval," applicants must have six years of experience as a TV weather forecaster, pass an AMS test, and submit videotapes of some on-camera appearances.

Forecasters like these probably do dream of perfection, as does National Weather Service professional Hallgren. Even if perfection ever becomes possible, however, I wonder if there might not be a question about its desirability.

If computers could predict our careers and our lives accurately, how many of us would dare to know about approaching poor health and date of death? All of that information would probably cause more anxiety than the world's population now endures.

If weather forecasting *never* becomes an exact science I won't mind. Perhaps the day will come when, at 5:15 a.m. on a cloudless morning, some computer can spit out with accuracy a forecast calling for a tornado to strike my neighborhood that afternoon. It might help me if I'm not insured: I can rush to some office and get appropriate coverage. Chances are, however, that the insurance company will have anticipated such action and will have clauses galore to protect itself.

The fact is, I'd rather not know for certain what the weather is going to be, although it would also be undesirable to return to the censorship days of World War II, when United States weather news was blacked out for fear of somehow aiding the enemy.

Perfect weather forecasting would take some of the interest, mystery, excitement, and anticipation out of living in Texas. How pleasant would it be to wake up in the midst of a five-year drought and *know* there couldn't be any significant rainfall expected for another two years? Give me Texas weather—some of it in small doses, if possible: the droughts, floods, winds, blizzards, tornadoes—and let me be able to hope for the best, whether future days prove to be hot and sunny, rainy and cold, or stormy.

Note

1. As is almost always the case among meteorologists, it seems, there is disagreement on this subject. A National Weather Service meteorologist in Waco, Doyle Casey, was recently quoted as saying that computers have made no difference except in speed of forecasts.

PART TWO

2

7. HOT AND SUNNY

It must have been in 1980 when Al Oliver, then a baseball player with the Texas Rangers, walked onto the field at Arlington Stadium in gathering dusk to play a summer night game. In areas farther north, where major league baseball originated and remained for many years, this would have meant, of course, that Oliver could have expected a cool temperature—or at least cooler temperatures than those recorded during the warm, muggy afternoons that can plague northern areas. But at Arlington, Texas, that evening Al Oliver walked into an oven. The thermometer hadn't yet gone below 100 degrees Fahrenheit in its evening slide.

"If hell is any hotter than this," Oliver was quoted as saying, "I'd better get my act together."

Texans, especially Ranger fans who shared the discomfort, could laugh at Oliver's funny remark about the heat. But another comment, this one a genuine history-maker, aroused great regional anger. While Philip Sheridan, a native of Albany, New York, was serving with the United States Army in Texas around the Civil War, he spoke one of the greatest blasphemies ever directed against the Lone Star State. "If I owned Texas and all hell," he said one torrid day, "I would rent out Texas and live in hell."[1]

Sheridan later compounded the hatred Texans thus had developed for him. In postbellum years, while acting as military governor of Louisiana and Texas, he instigated Reconstruction policies of such harshness that President Andrew Johnson removed him from office. As a result, Sheridan's popularity in Texas ranged from low to rockbottom. (There *is* a community in Texas called Sheridan, in western Colorado County, but it was named after a pioneer settler whose first name was Hugh.) Still,

Phil Sheridan later displayed diplomacy sufficient to defuse some of the hatred. It happened during an 1880 visit to Texas in the company of former president (and former general) Ulysses S. Grant.

On the night of March 24, 1880, Galvestonians honored Grant and Sheridan with a banquet in their famous old Tremont Hotel, despite the doubts that hosts had about the dinner's success, considering the backgrounds of their two guests.

General Sheridan came through handsomely. In a brief talk he explained that his notorious Texas-and-hell remark had come after completion of a hot, dusty stagecoach ride through the dry interior, far from Galveston. Then he added some words of praise about the state. Afterward, many hands slapped his back.

But the fact has survived Sheridan: Texas, in summers and not infrequently at other times, can be *hot*. The hottest temperatures are invariably registered inland, not along the coast, but stifling coastal humidity can negate that advantage unless there is shade and sea breeze to assure comfort.

Fourteen Texas towns with high readings of from 117 to 119 degrees are close behind Seymour (with that 120-degree reading) in the dubious contest for hottest place in the state. Some have recorded their maximum temperatures more than once. In the 117-to-119-degree category are Big Spring, Clarendon, Graham, McKinney, Memphis, Mount Pleasant, Pecos, Presidio, Quanah, Throckmorton, Tilden, Vernon, Weatherford, and Wichita Falls. Sunshine, too, is most often observed away from the coast, where the same humidity that makes it seem hotter also can contribute to cloud formation. The far western area known as the Trans-Pecos can boast (if they want to boast about this) of having the most days of sunshine: only one out of every six is cloudy, except for winter, when the number of cloudy days doubles, approximately. But this still surpasses by far the number of cloudless winter days throughout much of the rest of the state, where fewer than half can be thus described. The season that is generally the *least* cloudy in all of Texas is autumn.

I recall the remarkable sunshine and heat of one winter day in the late 1950s when I moved to Austin for a job as assistant to the director of the University of Texas Press. The early-February day before I commenced that employment, a thermometer at the house my wife and I had just bought registered 92 degrees. In Texas this won't prove unbelievable, but I doubt that any resident of Buffalo, New York, who might read this would fail to show some surprise.

Another reason, in addition to the previously mentioned summer solstice and ensuing heat-lag, for the prolonged and uncomfortable warmth throughout most of Texas during summer months is the arrival from the tropics of a "Bermuda high," an upper-atmosphere ridge of high pressure that lingers and, as is usual with high pressure, brings fair weather in large doses. Thereafter, the sun's rays can stream down upon Texans with very little moisture in many areas of the state to soak up the heat. But moisture mixing with heat usually does nothing whatsoever for personal comfort. If I have to endure extreme summer heat (and in Texas that can be taken for granted), let it be as dry as possible, even though I am a fan of slow, cool rains.

The two types of heat stress, wet and dry, can both pose dangers. In dry heat the stress comes from large doses of convective

HOT AND SUNNY

Sunshine recorder, a glass tube containing mercury, used by J. H. Jarboe, San Antonio's "weather prophet."

and radiant heat; in humid heat it comes from restriction of the evaporation of sweat. (There are, of course, artificial heat stresses to be found in some industrial jobs, but they don't have a place in this book.)

During conversations in earlier years, when I was innocent about the humdrum possibilities of such talk, it was the humidity I often blamed for my discomfort—as did and still do thousands of others who don't know the specific reasons for this. A major reason for the relative comfort of dry heat is that the less humidity there is, the easier it is for perspiration to evaporate and to cool skin and blood. That's why desert heat is more easily endured (by most individuals, anyway) than humid heat: body moisture passes into the dry atmosphere with greater ease.

When, on a warm day, the thermometer registers 98.6 degrees Fahrenheit or below (normal temperature for a human body) and the humidity is relatively low, body heat is even less a problem, since heat energy is always passed from higher to lower temperatures.

The human body does this through the cooling of blood. While traveling through capillaries near the outer skin, blood gives off some heat into the atmosphere, keeping body temperature around that level of 98.6. But when atmospheric temperature is higher than 98.6, there are obvious difficulties. The body depends on sweat evaporation for temperature control, but sweat evaporation then is more difficult.

Thus the warmer the temperature and the higher the humidity, the more uncomfortable the day becomes in the great outdoors, even for healthy young Texans; and actual danger can develop for others with circulatory problems. Hot weather forces the heart to pump more blood in an increased effort to cool the body in that remarkable way already described. High humidity combined with excessive heat results in that loss of cooling ability so that the body seeks to compensate through the respiratory system, which could result in such a loss of carbon dioxide (through overoxygenation of the blood by rapid or deep breathing) that hyperventilation results—dizziness, fainting, disorientation, or worse. The dangers of extreme Texas summers are not to be taken lightly, not even by the young and healthy. Attesting to this are the occasional news stories about athletes who have collapsed and died in the rigorous heat of late-summer football practices. I can personally attest to another danger, though usually it is a bit less severe: skin cancer, which appears to be more prevalent and more severe nearer the equator, where sunlight is the most intense, except that blacks, Orientals, Latins, and American Indians apparently have considerable immunity.

As a teenager, and even into later years, I lazed in the sun with impunity, as lightly clad as feasible. This served not only to get me out of a great deal of work, but also to enable me to avoid the deathly pallor that, alas, has been completely out of fashion for generations now.

Years later I wished I had settled for more pallor. When I was thirty-three the first trouble appeared: a red area on my sternum (which was on my chest, not where some of you readers might think it was). It began itching, and, as I swabbed it with alcohol more and more often to stop this, it became inflamed and ulcerated. It would not heal, no matter what I tried. Skin cancer. A doctor removed it with radiation treatment. Then he asked about my sunbathing habits, advised stopping altogether, and remarked that I could expect more skin cancer to develop as a result of exposure to the sun years earlier.

He was so right. More than a quarter of a century later I still return to skin specialists from time to time as new spots appear. This is an ailment one hundred times more prevalent in Texas than in northern states like Minnesota and Vermont, so Texans should be warned of another danger in regard to it. Some medications can make skin even more sensitive to the sun: sulfur, tetracyclines, and some tranquilizers, antibiotics, and birth-control pills. (Ask a doctor for advice.)

My kind of skin cancer is easily treatable, but another, rarer type can be deadly. Malignant melanoma should be cared for at once. If change of color or inflammation is noticeable around a mole, consult a doctor quickly. This is especially true for redheads and blondes; they are particularly vulnerable. As for the less dangerous types of skin cancer, fair-skinned blondes and blue-eyed individuals should beware of much sun exposure.

Skin cancer can sneak up on a person years after the original damage was done, as was true in my case. Other types of sun and heat problems can sneak up on people, too, but without so much delay. Some of these have already been mentioned. Heat cramps can result from loss of liquids and salt. Any such cramp is painful, but one in the stomach while swimming can easily result in drowning. Liquids, massage, and rest provide the best relief. Heat syncope can strike a person who has been exercising too long and too energetically in the hot sunshine or another warm place. That person will feel dizzy and might faint, and will have a weak pulse with a rapid heartbeat. Liquids and relaxation in a cool place will help with reviving.

There are more serious ailments that exhibit similar symptoms. Heat exhaustion usually occurs over a longer period of time, as a result of losing body fluids and salt. Victims will complain of headache, weakness, weariness, nausea, and disorientation. His or her skin will feel clammy and cool but will show pallor and perspiration. Bed rest in a cool place and cool nonalcoholic liquids will be needed for recuperation.

Any of these heat-related illnesses could result in heat stroke, or hyperthermia, which is always a serious matter. Heat stroke victims need medical care at once, especially if they are elderly or have heart ailments. At first they complain of headaches, dizziness, confusion, and nausea. Victims feel hot and dry, show a noticeable flush but no perspiration,[2] suffer soaring temperatures (sometimes to 107 degrees or even higher), and have a fast pulse. Medical help is needed urgently, but while waiting for it the victim should be undressed, put in a cool place, and sponged with water if possible.

These suggestions come with no guarantees from me; I am no more a medical doctor than I am a meteorologist. But I can speak with *some* authority about Texas torridity, having suffered from heat exhaustion at least once. Until I read a newspaper column recently by Sally Squires of the Newhouse News Service I had presumed I was somewhat unique in this regard, but in that column Dr. James Knochel of Dallas was quoted as saying, "Probably everybody in this country has had heat exhaustion . . . usually . . . during the early parts of a heat wave."

The trouble is, apparently, that one cannot really be certain. Heat exhaustion (but not the deadly serious heat stroke) can seem like a summer virus, with the victim complaining of headache, fever, muscle pain, backache, and nausea. These symptoms often disappear with the body's adaptation to the heat. If care isn't taken, exhaustion can lead to heat stroke. In that same column Dr. George Zenner of Houston spoke of the danger of overheating, particularly for aged individuals and for infants, for persons who are overweight or diabetic or show a history of heart disease, and for still another group that might not have realized the vulnerability: people taking medication that interferes with sweating. This medication can include antihistamines of the type found in common nasal decongestants, diuretics (which increase urination), and certain drugs used to control blood pressure, degenerative diseases, depression, and the like. Some of these drugs can also create another hot-weather danger: preventing a recognition of thirst. (Again, if you are concerned, ask a doctor.)

Any summertime resident of the Lone Star State can be sure there is another certainty in addition to death and taxes, and that's sweat. Dallas is reputed to be the fifth-hottest city in the United States. (Phoenix is first.) That eclipses any claim by Houston, which often seems even more uncomfortable because of its humidity. (*Dallas Times-Herald* columnist Molly Ivins once remarked that Houston has the city-spread of Los Angeles and the climate of Bombay.)

That must have been very troublesome in years now long gone. Old residents have recalled sleeping in summer on porches screened by mosquito netting, hoping to catch any wisps of a breeze, while insects buzzed around them. Surgeons operating with the supposed aid of three-foot fans would have nurses assigned to wipe away sweat from their faces, and whenever drops of perspiration fell on patients they would cease work temporarily while someone covered the contaminated area with drops of ether.[3]

Today's Houston has one great advantage, of course, over yesteryear's: air-conditioning—of homes, automobiles, offices, cattle auction barns, sometimes entire malls, and at least one athletic field, the Astrodome. And, of course, that's true of all Texas. Rare indeed is the new structure that is built without it.

Serious efforts to lower hot, humid temperatures began in the early decades of the nineteenth century, and maybe before that. Florida historian Raymond Arsenault credits a Florida doctor, John Gorrie, with developing a hospital cooling system in the 1830s: forcing air to blow over buckets of ice suspended from the ceiling of an Apalachicola hospital, to lower the body temperatures of malaria and yellow fever victims.[4]

In 1882 came the electric fan, but it did nothing for humidity. In 1924 theaters in Dallas (the Palace) and Houston (the Texan and the Iris) were the first in this region to be artificially cooled. In 1930 a Houston Cadillac owner purportedly devised the first air-conditioned automobile by installing a condensing unit in the trunk and a blower behind the driver's seat. And so on, into the 1980s, when Texas led all other states in air-conditioning. Wherever one wanders outdoors in Texas during the summer, however, he or she will usually find broiling middays. Some local chambers of commerce and some citizens seem as reluctant to admit this as they were in the days and months following Phil Sheridan's miserable stagecoach ride. For that reason I enjoyed reading another Dallas columnist, John Anders, who wrote in the *Morning News* July 10, 1984: "Summers in Dallas, Texas, are not meant to be shared; they were created to be endured. Texas summers are our shame and our sorrow, and we should have the decency to keep them to ourselves." Five weeks after Anders wrote that, many Republican delegates to the national convention were surely thinking similarly, even if few of them expressed publicly any negative opinions that would encourage spiteful Texans to vote Democratic. On the day most delegates arrived in Dallas—Monday, August 20—readings of 106 degrees and higher in the vicinity set records, and hospitals reported several cases of heat exhaustion, though nothing serious. "The only thing worse than this," a diplomatic delegate from Illinois remarked, "is Chicago in January."

Meanwhile, inside the Dallas Convention Center, people experienced uncomfortable chilling with thermometer readings of below 70 degrees, produced by 5,800 tons of air-conditioning. During preconvention work the mighty cooling system had produced a manmade "norther," with air chilled to 62 degrees. A jacket-clad security guard had compared it to Antarctica. "The other day, when I didn't have my coat, my nose was cold, my hands were cold, my feet were cold. It's got to be a big waste of money." A few days later a group of young demonstrators from outside the state lay down in a city street near the Convention Center, hoping to make things miserable for delegates, but they sprang to their feet soon after touching the burning pavement and went about other business.

During the convention, television newsman John Chancellor discussed air-conditioning and its effects on the summer Southwest. On an edition of the "NBC Nightly News," he remarked that Dallas "sits here, in the middle of nothing, looking like the Emerald City of Oz sprouting skyscrapers. And when you think about it, none of this would be possible were it not for something Americans take for granted—air-conditioning."

Cooling caused a profound change in the country, Chancellor said. There are skyscrapers that could not exist without it. Computers in them would not work without it, and "probably neither would the people." Because of this technological advance, the great migration of Americans into Sunbelt states began, bringing "voters who tend to favor Republicans over Democrats."

No wonder, Chancellor might have added, members of the Grand Old Party have become stereotyped as fat cats. In Houston the bill for air-conditioning alone came to $666 million in 1980. Only individuals with more than a little money can pay electric bills assuring complete comfort today.

Notes

1. It was indeed General Sheridan who said this, and not General William T. Sherman, who is sometimes credited (or discredited) with the remark.

2. Sweating stops just before a heat stroke, and body temperature rises alarmingly.

3. See Clifford Pugh, "How air conditioning tamed the beast called Houston," *Houston Post,* August 7, 1983.

4. Mentioned in a 1984 issue of the *Journal of Southern History*.

8. DRY

My greatest concern about dry weather in Texas antedates 1949, as will be seen in a later chapter, but some reading in old newspapers that year excited my interest. Conducting research for a thesis at the University of Texas, I came across this description of a Central Texas drought in the *Comanche* (Texas) *Chief,* September 18, 1879: "Not within the memory of the oldest inhabitant was water scarcer . . . than at the present time. With the exception of light and unsatisfactory showers, there has not been a copious rain since the 5th of May last. Many people in the country are hauling their drinking water a distance of five, eight and ten miles. Stock, as far as practicable, have been driven west, out of the country. Between Waco and Belton, forty miles, there are but two places where a man can get a drink of water."

That caught my eye immediately: two oases between Waco and Belton. In the summer of 1940 my parents had bought what probably was one of them: a ranch on Deer Creek, along which their 320 acres lay. The site was on what was then U.S. Highway 81, now Interstate 35. The purchase came during one of our summer visits from California, where we lived then. My parents, both native Texans, intended to (and did) retire there later, near their own parents.

I recall my father showing me his new ranch for the first time after he had bought it. Buffalo grass covered rolling prairies that dropped down toward the trickling, rock-bottomed creek, shaded by tall pecan trees. He led me east along Deer Creek, past a windmill and in the direction of what I knew to be a bridge on U.S. 81, although the automobiles couldn't be seen because of thick greenery along the creek. I could *hear* them. They were only an occasional noise; that was a quieter, not-so-crowded time. At a location where the creek had, through the years, formed a natural rock dam, he stopped and told me some history passed on by the former owner, whose family had acquired this and much surrounding land many years ago. The water hole behind the dam, helped by some natural springs, had never been dry in the former owner's memory, he said, and the man said his father had told him the same thing.

Even in those much-younger days I was cautious about believing any stated adequacy of a Texas water supply. I don't remember telling my father outright that I doubted the claim, but I did. He didn't, since he had come to know the former owner through the land purchase and had concluded that the man was honest. For solid old-time Texans, who seemed to be good judges of character, a man's word was sufficient documentation. After my father and I had stared for a few moments at the water hole and the catfish that could be seen swimming slowly and apparently asking for a hook, he pointed to what looked like a ravine cutting at a right angle into the opposite bank. What appeared to be a long-unused road led down to the level of the natural rock dam, then across it, and up through another ravine on our side of the creek. My father said that the ravines had eroded from stagecoach ruts dating back to the earliest days of settlement in this part of Texas. The former owner had told him that, too.

My parents died in 1976 and left the ranch to me. By that time I had happened upon information (already known to many others, but not to me) indicating that IH 35, U.S. 81, and earlier highways along that route generally followed an old stage road used long ago. That same information had informed me that at a site near the northern boundary of Bell County there once had been a crossing, near a water hole, where stood a blacksmith's shop and one or two other businesses.

In the late 1970s our son, Don, poked around the area with a metal detector and uncovered horseshoes, nails, and other items. So now I also know where, beyond doubt, could have been found, in the dry year of 1879, one of only two places between Waco and Belton where a man could get a drink of water. As I write this in the dry summer of 1984, things are even worse in some ways. Years earlier, in 1953, our historic water hole had gone dry for the first time, a victim of what I have since read was *really* the worst drought Texas ever had. By December 1956, 244 of 254 Texas counties constituted a government-declared disaster area. Dallas officials and authorities in some other Texas cities watched municipal water supplies fall to alarmingly low levels, and in Dallas, at least, the only additional water obtainable in quantity contained too much salt for residents on low-sodium diets. A few freshwater wells were dug around the city, historian A. C. Greene wrote recently, and people requiring low sodium could get drinking water free there, as long as they took it away in containers holding only one gallon.

It was that same drought that dried our formerly reliable water hole for the first time we knew about, and it has never been quite the same since. After a comeback following the drought of the 1950s it went dry again, and it has alternated between wet and dry in recent years as the rains come and go.

In 1879 a dry water hole at what is now our ranch on Deer Creek might have proved

After a dust storm of 45 minutes duration at Stratford, Texas. April 14, 1935.

fatal to men, women, and beasts, but these days inhabitants in the vicinity enjoy certain benefits not available to the old-timers. Pipes carry water from wells and lakes directly to their homesteads. This has resulted in some new problems. Many years ago rural residents were aware of the frequent inadequacies of rainfall. Today, some of them tend to take water supplies for granted, not realizing the limitations to be found beyond the faucets. Texas growth—in fact, all Sunbelt growth—is at the mercy of the water supply, and the energy supply, too. A fifteen-minute power failure at that 1984 Republican National Convention in Dallas proved terrifying to people who realized the potential. With temperatures of more than 100 degrees outside, they could visualize what would soon happen to a Dallas Convention Center that lost its air-conditioning for long. But the danger passed.

Summertime power failure lasting only a few hours can cripple Sunbelt cities now as never before, but presumably the failure would be quickly rectified. A drying water supply is another matter, and considering the reputation that Texas has had for frequent drought, it is difficult to understand why there has not been more of a push to try to solve water problems.

Drought and desertification (a process by which marginal lands become desert wastelands) can be intertwined, as has been shown in Africa. Worse, some scientists theorize today that desertification feeds upon itself. As vegetation vanishes, the land surface reflects more sunlight, creating more dry air that sinks toward the land surface and compounds aridity. Under such conditions even a mesquite tree, the thorny Texas infestation that *can* be beneficial by producing beans that cattle can eat when other fodder has disappeared, has its life limits, although it sends roots down as far as forty feet to suck up any moisture available in the earth.

A full-blown drought, however, is not needed to bring on desertification. Humankind can help do it with little aid from nature, by overcultivating and by cutting down forests. Water and wind erosion can follow, and the sun-baked soil will eventually be unable to absorb and store water.

Faulty agricultural use of the land cannot be blamed at all for many southwestern droughts. Two of many clippings in my files corroborate this. The first one concerns some 1957 studies of dust layers in a Palo Duro Canyon cave by members of the Panhandle-Plains Historical Society. Long before soil in the area was tilled, evidence shows, the land suffered extended droughts of perhaps even greater severity than in modern times. Paleontologist Jack T. Hughes, who directed the excavation, speculated that the several dust layers uncovered in the cave provided a calendar dating back more than a thousand years and, through the dust layers, gave excellent evidence of some awful droughts that sent inhabitants fleeing to other climes.

Three years before that study came one, in 1954, by University of Texas archaeologist Alex Krieger, using tree rings, old Spanish diaries, peat bogs, diggings at old Indian camps, and radioactivity measurements to get details about Texas and southwestern United States weather and climate. Krieger found that short droughts lasting two or three years have been common, and that, for the past thousand years or more, droughts lasting up to fifteen or even thirty years have come along every other century.

A thirty-year drought these days would motivate even the weekend picknickers and golfers to wish for rain any Saturday or Sunday they could get it.

Many other studies have attested to the likelihood of another disastrous drought in Texas. Still, there is reluctance about expanding the water supply. One reason is, of course, that some people would have to make inordinate sacrifices. New lakes cannot be planned without the condemnation of land and buildings that have been in family hands for many years. Something will need to be done, however, and only government has the power to do it. The late Speaker of the House Sam Rayburn knew how to manage it. Years ago, when an Oklahoma governor and some Oklahoma congressmen voiced opposition to his plan to create Lake Texoma on the Red River between Texas and Oklahoma, claiming that too much Oklahoma farmland would be flooded, Rayburn sent word that since the state desired to protect its soil, it would get no more federal funds for any other such project.

In that instance Rayburn was working largely for his own Texas district. The proposed lake would give it cash inflow as a recreational area, among other benefits. But Rayburn used the pressure that will no doubt be necessary for doing much in the way of increasing Texas, and southwestern, water supplies. Lake Texoma ranks now as the tenth-largest manmade lake in the United States.

It seems, during the droughty time I happen to be writing this, that Texas could use at least ten more Lake Texomas spotted around the state, with emphasis on areas where the rainfall is least. Texas might be having floods as this is being read, but droughts will never in the foreseeable future be plagues entirely of the past.

West Texas has been especially vulnerable. Military records kept at frontier forts around the midnineteenth century show continual drought tendencies and modest annual rainfall amounts. Both are still expected these days, even in "normal" years. But those early frontier years came before the general advent of windmills, wells, artificial lakes, and dry-farming methods. A drought in those days could totally ruin life in the region.

In 1885–87, and particularly in the summer of 1886, a disastrous dry spell, thoroughly recorded by suffering residents, held much of inland Texas in a deadly grip, causing crop failures and great loss of cattle. During the entire year of 1886 Midland received only 7.3 inches of rain.

(Typically, however, another part of Texas—Brownsville—endured excessive rainfall that same year: a total of 60.1 inches, including more than 22 inches in one day, September 22. And an August hurricane, one of several large tropical storms that dumped huge amounts of water on the Texas coast that year, raked, flooded, and demolished the once-important port town of Indianola; it was never rebuilt.)

The 1885–87 drought was so bad that it became a historical turning point for people who lived through it, as did the Civil War and World Wars I and II. For as long as survivors lived afterward, even well into the twentieth century, there were people who understood the term "before the drought" and "after the drought." In the 1920s W. C. Holden of Lubbock, still an active historian as this page is written, researched and wrote about it in an article entitled "West Texas Drouths" in the *Southwestern His-*

Pig tracks. Farm animals wander through the sand, following a Panhandle dust storm.

torical Quarterly (October 1928).

It was a drought truly to be remembered by a people who tended always to exaggerate the severity of rainless periods. From those several years of extended aridity West Texas acquired its national reputation—one still widely known because of subsequent weather—of being a perennial desert ravaged by hot winds (which is not, of course, a true description). Nature's viciousness in 1885–87 forced people to evacuate the area in a fallback equal to the 1865–66 flight from the frontier of people besieged by renewed Indian attacks. By late May 1886, residents of Cisco were completely out of water. The Houston and Texas Central Railroad began running a water car from Albany, thirty miles northwest, and saved the town from total abandonment, although residents had to pay for their nourishment by the bucket and by the barrel. Cattlemen moved stock to any location where water might be available, then camped there with the animals. By June and July, immigration, already slowed to a trickle in the drought area, ceased completely, and people began to move out en masse—the poor first, of course: their credit had been used up. Industry, trade, and civic work came to a halt. Cattle died by the herds; water for the animals could no longer be located.

Other gaunt cattle, driven to fringe areas of the severest drought, hung on, but they plagued farmers who had been able to stay there and work their stunted crops. The cattle would devour the scant pasturage and drink up whatever water they could find, then break through fences into fields and eat whatever thin crops they came across there. Some skimpy showers eventually helped prevent a bloody outbreak between ranchers and farmers; cattlemen were able to turn their stock away from the areas under cultivation.

Autumn and winter passed with no relief despite increasingly devout prayers for rain by the decreasing populace. A newspaper in Mason reproduced the text of a chalked note left on the door of a deserted house two counties away, in Blanco County: "250 miles to nearest post office; 100 miles to wood; 20 miles to water; 6 inches to hell. God bless our home! Gone to live with wife's folks."

Roads manifested a westward expansion in reverse. Covered wagons traveled eastward, loaded with family possessions, accompanied by miserable refugees, and drawn by bony animals. As many as half the inhabitants of the stricken countryside fled, sometimes leaving whole communities deserted.

But a number of stay-behinds still managed to find enough water somehow to quench their own thirsts and to laugh at their plight in a manner peculiar to hardy Americans who, throughout history, have endured unspeakable weather and frontier dangers, great economic depression, and wars. Maybe they found laughter to be their last solace, thus unconsciously corroborating Sigmund Freud, who declared that laughter is one of the best human defenses against suffering and pain. Whatever the reason, a number of newspaper editors, especially, tried to be funny in an unsophisticated nineteenth-century way, and they contributed some tall Texas-weather tales recorded by the Lubbock historian W. C. Holden.

"The weather has been so dry here for the past three weeks that the wells are empty and the fish in the creeks are carrying [toadstools] for parasols to keep the sun from burning their backs. Water is getting so scarce that Baptists and Campbellites are beginning to favor baptism by sprinkling, and they have quit turning up their noses at Presbyterians. . . . A prominent prohibitionist has ordered a case of beer from Decatur . . . as . . . lather to shave himself" (*Taylor County News*).

"Monday morning an individual was seen in town with a long slicker under his arm, and it naturally created intense indignation. . . . S. C. Hines . . . swore out a warrant against him, charging him with . . . carrying a slicker against the peace and dignity, dampness and future prospects of rain" (*Albany News*).

But there never have been and probably never will be many aspects of a Texas-sized drought that are genuinely funny. Two small exceptions come to mind immediately, but the first one would scarcely be regarded as hilarious by persons concerned. In recent years of aridity, Lubbock city officials began cutting through the clay floors of playa basins—small natural lakes serving as reservoirs—to enhance the city water supply. But this apparently resulted in the development of leaks from the playas into the aquifer below, causing the water level there to rise beneath the campus of Texas Tech University and flood basements. Worse, the level has been reported "precariously close" to the floor of Jones Stadium; and no one knows how drastic might be the effect on students and alumni of a major American university forced to give up football. (Ironically, Texas Tech scientists have devoted years of research to finding ways of artificially recharging underground water supplies in the area.)

Another humorous exception: early in 1955 the Texas Society of Professional Engineers presented a plan for water conservation to Governor Allan Shivers. The story ran in many Texas newspapers, and there it created some problems because of a few style sheets (which state individual newspaper preferences for punctuation, spelling, and so on). One common spelling preference was "drouth" instead of "drought." The president of TSPE that year was San Antonio resident Frank Drought, and at least one newspaper (the *Tyler Morning Telegraph*) called him "Mr. Drouth" in the story it ran. But the 1885–87 drought provided few such laughs. As is common in most catastrophic situations, drought victims who remained in the stricken area and sought to survive displayed a remarkable spirit of unity and mutual aid. Contributions came from other Texans and from residents outside the state, but, according to W. C. Holden, "The greatest sacrifices in the way of helping others came from within the [drought] area itself. . . . A notable example occurred in Runnels County. . . . The citizens of Ballinger, although they were having a hard enough time themselves, [sent] a wagon load of flour, meal and bacon. Within a month twelve more wagon loads had been sent."

Finally, in May 1887, a series of spring rains eased the drought throughout most of the area, although on June 5 Abilene residents experienced a lingering reminder: a sandstorm whipped by winds exceeding fifty miles an hour that reduced visibility in the town and surrounding areas to less than fifty feet. It wouldn't be the last time Texans endured such an irony of weather.

9. WINDY

Who hath seen the wind? I have, that's who—at least as much of it as I care to see. For the most part it was when I was growing up in the West Texas city of Abilene.

It wasn't *really* the wind I saw, of course, but the effects of it. And in those days of the early 1930s the effects could be downright dismaying, even though Abilene lay on the fringe of the real Dust Bowl. Clouds of dust and sand would roll in from the west, but sometimes in the early years of those troubles the clouds looked, from a distance, more like promises of rain to my inexperienced eyes. Before long my nose and throat, as well as my eyes, learned to tell the difference easily enough.

During those days, I have been told, I suffered from some kind of dust pneumonia—apparently a mild type that comes from inhaling wholesome natural dust (it seemed more like sand to me) and not the truly dangerous stuff like coal dust, asbestos, carbon, and the like.

Just now I have reread the "dust" entry in the old and reliable medical encyclopedia that I have referred to throughout the years since its publication around the time Hippocrates was practicing medicine, and I see that it begins, "Dust is disastrous to human beings." That is indeed how I remember the West Texas days of my youth. But, incredibly, the last paragraph of that entry indicates that the immediate hazard to human health from inhaling those parts of West Texas, New Mexico, and Oklahoma that drifted into my nostrils was not particularly great. "Most of the dust that gets into the nose and throat . . . may be washed out, blown out, or coughed out. . . . The worst effects of such dust are the mental effects and the occasional association of the dust storm with secondary pneumonia."

The pneumonia I had, however, seemed more of a primary type to me; and I am sure

I have never completely washed, blown, or coughed out all that dust. That would have been impossible. Somewhere deep within me there remains a topsoil residue from somebody's long-since foreclosed farm, probably looking like the grit that remains on the floor of a freshwater shower after beach-bathers have washed themselves and their swimsuits, dressed, and gone home.

The dust, or sand, or whatever it was, seeped through windows and doors until we could see the marks of our feet or shoes on the hardwood floors. During storms I slept with a wet cloth over my nose, but with little comfort even then.

And this was on the fringe of the Dust Bowl. Things were worse in the Texas Panhandle and in areas east, north, and west of there. Very descriptive of those years is a book in my library that I have just looked at again: *The Dust Bowl: An Agricultural and Social History,* by R. Douglas Hurt (Chicago: Nelson-Hall, 1981).

When a "black blizzard" struck, living became miserable indeed, Hurt wrote. Boiling dust clouds extending fifteen hundred miles horizontally, reaching upward to (on at least one occasion) fifteen thousand feet, and whipped by winds of fifty or sixty miles an hour would turn day into night for as long as thirty-six hours. Lights would flick on in automobiles, stores, offices, and residences, but the raging storm would darken their shine.

On railroads, locomotive engineers sometimes could not see warning signals and would ram other trains or would tear into vehicles attempting to cross the tracks. Sometimes they couldn't even see the depots and would pass them by. On several occasions crewmen walked ahead of a creeping train to try to see, in the daytime darkness, whether dust had piled high enough on the tracks to threaten derailment. Sometimes they used snow plows to remove the dust. Once, when drifting soil derailed a train, a rescue crew sent out to help became stranded for two days.

The worst storms also paralyzed city traffic for hours at a time; piled dust in some attics deep enough to cause collapse; brought death in one area to at least six persons and illness to more than one hundred others in a period of two weeks; sent so much powdery soil into a college gymnasium that a basketball game had to be called off; once brought a fifteen-minute period of darkness described as absolute to one community; sandblasted paint from automobiles; and sometimes put a frosty appearance on windshields, necessitating replacement. Plains soil blew far and wide. One day in Austin, outside the Dust Bowl proper, state senators wore surgical masks to a session—and not because of any legislative stink. Dust covered far-distant pastures, too, and coated grass so thoroughly that cattle couldn't eat it. In May 1934, wind scooped up an estimated 300 million tons of plains soil and carried it to the eastern United States—as far away as New York City, and even beyond. Five hundred miles out in the Atlantic Ocean, crewmen aboard ships noticed the settling dust. Not all of it had come from Texas, of course, but Panhandle farmers who had been seeing their topsoil disappear probably thought so.

Under such miserable circumstances Texans—and, in fact, Americans everywhere—can become tenacious. Many persons were forced to leave the land, of course, and I saw them when my parents and I moved, for more pleasant reasons, from Abilene to

WINDY

A dust storm in West Texas.

An approaching dust storm in the Middle West. The Dust Bowl of the 1930s extended outward to encompass the Texas Panhandle, Western Kansas, and large parts of eastern Colorado and New Mexico.

California in 1937. Their old cars, almost always piled high with children and possessions, and with bare mattresses tied down atop each vehicle, crowded famed old U.S. Highway 66, which we traveled that year. I still remember the automobiles stopped along the highway with radiators steaming or a tire as flat as the plains those poor people were leaving; but to my youthful credit I never snapped a photograph of those unfortunate folk, though I wish now I hadn't been so sensitive. The children I recall seeing would have made good camera subjects. They were not crying, no matter how the adults felt about things; they seemed, instead, alive with excitement about their adventure.

Not everyone moved off the land, especially if he or she owned part of it, and in some recent reading on the subject I see that optimism and an insurmountable sense of humor must have been largely responsible for the maintenance of residence, life, and sanity.

Rain would come again someday, people knew, even if it seemed later to take several fiscal eternities for this to happen. What little rain did fall in those years didn't wet the ground for very long; a forty-mile-an-hour wind can dry things sixteen times as effectively as a ten-mile-an-hour breeze. Meanwhile, however, many people could somehow laugh at their predicament. One such time came when self-proclaimed rainmaker Tex Thornton tried unsuccessfully to bring rain to an area around the Texas Panhandle town of Dalhart. The steady booming of all those big guns in World War I had brought plentiful rains to France, it was said around Dalhart, where people obviously didn't know much about the French climate. Explosives detonated in Dalhart on a day with low clouds and moderate temperatures would surely have the same effect, Thornton and his followers thought.

Area residents contributed $300 toward the purchase of explosives for use by the rainmaker, who scheduled his attempt for the first day of May, 1935, at Rita Blanca Lake nearby. Meanwhile, optimistic merchants scheduled a street dance for later that day: a celebration in the rain, they hoped.

Thornton planned to send his explosives skyward, in balloons, for detonation by time fuses at previously selected heights. A large crowd gathered to watch results; included were newsreel photographers from distant cities. Just before the blasting was to begin, a dust storm roared in, precluding the use of balloons. Spectators scurried for the safety of their automobiles. Once inside, they rolled up windows and strained to watch Thornton.

The man appeared undeterred. He decided to bury his explosives in the sand and set them off there. The ensuing din hurled more soil into the air, but it brought no rain. Whatever happened to the scheduled dance went unrecorded as far as I can tell, but presuming the newly arrived dust storm was a typical one, nobody would have lingered very long outside on a Dalhart street that evening.

That story can serve to introduce some Dust Bowl humor rounded up from various sources and retold by author Hurt in his interesting book. The stories are no doubt funnier now than they were at the time.

One story came from Kansas, but it could have pertained to Texas storms, too: "Lady Godiva could [have ridden] through the streets without even the horse seeing her," a journalist wrote about one black blizzard in his area.

Many other stories made the rounds. A raindrop struck one man on the head; it took two buckets of sand thrown in his face to revive him. . . . Post holes left standing above ground after the wind had blown away the sand around them were carried by farmers to their barns and stored there for future use. . . . Frogs captured and placed in water tanks drowned because they had never learned to swim. And so on. But there were true stories, too (or at least stories told as true by eyewitnesses), and the real ones could sometimes be as incredible as the imagined, Texas weather being what it is.

A West Texan remembers an experience recounted by her father, who was a Knight Templar and wore that "outrageous" costume of feathers and an elaborate sword on special occasions. One Sunday the members of his group were to serve as casket bearers and afterward were to raise their swords over the open grave of a fallen colleague. By the time the ceremony began, a high wind had begun to blow. The knights, swords raised, stood beside the grave, and all of them were blown in, feathers and all.

The Dust Bowl centered in the Oklahoma Panhandle but extended outward to encompass the Texas Panhandle, western Kansas, and large parts of eastern Colorado and New Mexico. Nevertheless, residents far removed from that specified area had good reason for saying they lived in the Dust Bowl, too, as black blizzards sent their topsoil hurtling eastward through the dim atmosphere.

Any place on earth with wind and drought and not many soil-embedded rocks could become a dust bowl, it seems, but the area that looms so large in our memory of the 1930s was particularly vulnerable. This was because an area east of the Rocky Mountains—an expansive plain largely devoid of trees, creeks, and ravines—had been farmed or ranched to a point of soil exhaustion, and it lay in an area of notable extremes of climate.

Those characteristics contributed to the advent of our Dust Bowl of the 1930s. More immediate causes were a gripping drought; high winds that were never unknown in the region but that, with the acute dryness everywhere, came to possess catastrophic power over the soil; and an increasingly barren landscape made up of crumbly and sandy soil that melting ice and snow had washed down from the mountains.

Dust bowl soils generally have a characteristic known as "flocculation," meaning they collect in clusters like fish eggs. Overcultivation combined with drought can break down this flocculation, and the soil becomes a fine powder that offers even less resistance to the wind but prevents normal penetration of moisture even by the few rains that come. Only a wet period extending over several years can regenerate flocculation.

Although the 1930s cataclysm did not have its actual origin in Texas, it affected much of the state and, as I have said, sent dust and sand up my nose and inside me so that some of it certainly remains in my craw to this day. Eventually, windbreaks like newly planted trees and new plowing techniques helped to mitigate Dust Bowl losses, but in the end only nature's return to normal rainfall solved the problem. Twenty years later, in the 1950s, the dust struck again, in the same general area; and with the onset of future serious drought it can surely threaten anew.

But the Dust Bowl owned no monopoly on the wind that can buffet the High Plains and other Texas areas. Obviously, wind can also come with drenching Gulf hurricanes, with heat-induced dust devils, with thunderstorms and tornadoes, with assorted weather fronts, and in other ways. The winds that can blow with unrestrained velocity across flat, treeless areas of Texas are mostly malevolent. In the summer, when abetted by a withering sun, they can bring quick disaster to farmers in several ways. For one, they help cause extreme dryness in the soil. Because of evaporation rates during an April-to-September growing season, the area around San Antonio needs sixteen inches more rainfall than land around Williston, North Dakota, to get an equal wetting. When those winds blow with furnace-like heat, as they sometimes do in July and August and adjacent months, they can devastate green fields of corn in a day or two. And when they roar in with a blinding, snowy blizzard they can kill people who are caught only short distances from their homes or automobiles. Winds blowing across open areas of Texas can average ten or twelve miles an hour, and sometimes more than that in the Panhandle. Those averages exceed that of Chicago, known as the "windy city." The windiest areas of Texas posed psychological problems for settlers from the very first, especially (it was said) for women. In 1925 a native Texas woman incorporated this into a novel. Dorothy Scarborough, an 1896 graduate of Baylor who later studied at Oxford and taught at Columbia University, published *The Wind,* a story of tragedy wrought in the soul of a sensitive woman.

At other times the wind can be delicious, especially when it blows gently on relaxing porch-sitters (of whom not many are left in these days of air-conditioning) on a warm summer evening. And along the Gulf of Mexico day-to-day breezes are, fortunately for people outdoors, often strongest in the muggy summer months. Land absorbs less heat than does water, causing the air above land to become hotter, then to expand and become lighter, and to rise rapidly on days when the sun's intense heat warms everything. Cooler, thus heavier, air from above the water then moves toward land to take the place of rising air, providing a sea breeze and a great deal of comfort, especially to persons who have the added advantage of a shade. At night the opposite happens, because land cools more rapidly than water.

At other times along the coast, and particularly around Galveston Island, high wind blowing seaward can have an additional effect. Strong northwesterly winds can keep tides abnormally low and, when powerful enough, can push enough water out of Galveston Bay back into the Gulf of Mexico to leave hundreds of boats grounded.

This has happened many times, including late February 1984. Some residents said they had never seen the bay water so low, and a few of them took pictures to record the sight. Oyster reefs showed themselves; deep-driven barnacle-covered pilings became exposed for the first time in years; the end of the boat ramp at the Galveston Yacht Basin Bait Camp, in the lower bay, stood two feet from the nearest water; and several miles of new beach appeared temporarily. Surprisingly, in this era of the fast buck, no con artists appeared to try to capitalize on the sale of this real estate. Ships waiting offshore to proceed to berths were forced to wait even longer because of low

WINDY

Storm-tossed sailboats in coastal waters.

channel water. Down the coast at Corpus Christi, low water left some large tugboats and a few other vessels grounded temporarily.

On the other side of Galveston Island from the bay, where Gulf whitecaps usually die with gentle sighs along the beach, high wind can change the scene, too. Strong breezes from the north can expand the beach. In 1956 a well-known Galveston old-timer and survivor of the 1900 hurricane, Walter Grover, told me, "[When] we had a north wind you could walk halfway to Cuba."

But on the morning of September 8, 1900, he discovered a remarkable exception. A 15-mile-an-hour northwest wind was blowing against unusually heavy swells rolling in from the southeast, and instead of causing a low tide, the wind was losing to the swells, which began flooding Galveston to a distance of several blocks inland. Later that day, when the wind changed direction (as it will do in a hurricane) and roared with 100-mile-an-hour fury from the southeast, it helped hurl the water over the entire area until every part of Galveston Island lay under the Gulf. From six to eight thousand persons died along the coast that weekend in what remains the worst recorded weather disaster ever to strike the North American continent. A detailed factual account appears in *A Weekend in September,* by John Edward Weems (College Station: Texas A&M University Press, 1980; reprint of 1957 edition). That storm led to the construction of the seawall there, and Galveston has not endured a comparable tragedy, although the city has suffered hurricane loss since then.

Hurricanes (to be mentioned again in a later chapter) come to the Texas coast only occasionally, however, and after they pass, the sea breezes take over again. But high wind can hit elsewhere in Texas with greater frequency. In any spring the High Plains of Texas can become one of the windiest regions in the United States, with "average" three-month speeds in excess of seventeen miles an hour. Gusts or sustained winds reach sixty miles an hour or more. The spring average might not seem so high until one realizes that the annual "average" of the windiest city in the United States—Dodge City, Kansas—is fourteen miles an hour. Other areas of notable windiness exist in Texas. Vacationers driving motor homes, house trailers, or other high-profile vehicles along U.S. Highway 62/180 between El Paso and Carlsbad Caverns National Park would be wise to tune in to radio weather forecasts; winds that tear through Guadalupe Pass along that route have been clocked at more than 110 miles an hour and have overturned eighteen-wheel trucks. When forecasts indicate the possibility of this danger, high-wind warnings are issued, and motorists are advised to use other routes.

In the spring of 1983 winds gusting to hurricane force (seventy-five miles an hour) near Rankin, fifty-five miles south of Midland, piled tumbleweeds to a depth of ten feet, blocking Texas Highway 349 and burying one car. Winds of such high velocities are not always needed to inflict damage in Texas and on Texans. The southeasterly wind that can be said to prevail over Texas takes its toll with the passing years—as it did with my grandfather's barn full of that penciled weather data. That same southeasterly wind worked on the century-old Red River County Courthouse in Clarksville so that in the early 1980s its clock tower showed a noticeable lean to the northwest, causing consternation to employees whose offices immediately below stood in danger of being battered by a collapse. Some persons refused to continue working in the building until repairs were made.

The Leaning Tower of Clarksville and tumbleweeds, hurricanes, and dust bowls notwithstanding, Texas wind doesn't blow ill all the time. I can recall the pleasure afforded by those summertime breezes in the days before air-conditioning. And there are other wind benefits.

For one, there's the windmill, which opened new areas of Texas to settlement. Earlier, the need for water forced immigrants generally to settle areas where a surface supply was available. The old type of windmill required some tinkering and alteration before proving satisfactory for pumping underground water in Texas. The original idea came from Europe, where people used windmills to grind meal and to provide other power. These contraptions were useless for pumping Texas water because of their large size and their need for constant attention as the wind shifted and changed velocity.

A few years before the Civil War a man in Connecticut solved the problem. He designed a wheel with wood slats leading from the center and set at angles to the wind, so that centrifugal force would slow the whirling during gales, and he put a "tail" on this wheel to make it always face the wind. A simple four-legged wood tower built over a well held the self-regulating machinery, which in its earliest days still required grease applied twice a week by a range rider.

Thus the same winds that can pommel Texas also bring forth water for cattle and gardens, and give men and women life, employment, and sometimes even prosperity. So much water has been pumped by this method and by others that the underground level is showing signs of exhaustion, but the same winds can possibly be of future benefit in other ways. The earliest windmills were used to provide power, and it might be that vast improvements on that old idea can result in producing energy. Texas has about as much wind as it has oil, and its depletion is scarcely noticeable, especially on a spring day.

10. COLD

New Englanders and other people who live in frigid winter localities far from Texas rarely give the state credit for having any cold weather. But as Texans know respectfully, cold weather is not unknown in the state, even at its southernmost tip; and extremely cold temperatures that would satisfy even an Alaskan Eskimo can, on infrequent occasions, take hold of the entire region in a matter of hours, by riding in on the back of an arctic blizzard that has stampeded out of some nebulous corral in the Far North.

Consider the aforementioned blizzard of February 11–13, 1899. That phenomenon howled into the state a quarter of a century before I did, but I can remember hearing my sky-scanning grandfather and a few other old-timers talking about it with an awe manifested by their voice inflections.

I have also read about that blizzard and can appreciate the awe, especially considering the absence of residential central heating (or in some cases, any heating at all, except for a solitary wood or kerosene stove in a flimsy frame house) and the absence of other twentieth-century improvements, both urban and rural, including better weather forecasting and dissemination.

Apparently few persons foresaw the arrival of that blizzard, although the preceding days had been colder than usual. Two polar fronts had already surged into Texas in February, on the fourth and the eighth, penetrating all the way to Brownsville. Snow had fallen in many areas of northern Texas and had blanketed the ground there, chilling things further for the arrival of subsequent fronts. On the eleventh, the third polar front arrived, bringing with it those record low temperatures already mentioned (like minus 23 degrees at Tulia) and eventually affecting people and things all the way to southern Mexico and El Salvador.

Ken Foree, an outdoors editor for the *Dallas Morning News,* seemed to enjoy reading old newspapers that described this bitterly cold wave, talking with eyewitnesses, and writing about it. Foree wrote several stories on the subject over the years, usually close to the anniversary date, and he had an affinity for beginning with a remark like, "So you think it's been cold?" Then he would hark back to that freezing February of 1899, when people had a real reason to complain.

By February 9, a Thursday, the ground at Amarillo had been snow-covered for two months, and—250 miles farther south—Odessa had been chilled by eighteen days of cold north winds. For eighteen days cold weather had also gripped Dallas, but that February 9 the mercury rose to a rather comfortable 45 degrees. The weather forecast out of Washington, D.C., for the following day called for colder temperatures in Dallas, with rain or snow likely.

This forecast proved during the next two days to be a great understatement. As the cold moved in from the Arctic, "near Coleman in West Texas a fog . . . came rolling in from the northwest, the icy fingers of the polar front reaching under the warmer air and condensing the moisture. That rolling fog felt like stepping into a refrigerated vault and farmer Jesse McGregor's household, which included a son, Stuart [later editor of the *Texas Almanac*], then seven . . . began scurrying around like a ship's crew at sea battening down for a storm. . . .

"In Gainesville a Negro boy of seventeen, Garrett Turner [later athletic trainer at SMU] . . . was working at the compress in season, going to school and hunting.

"Early [in the] morning of Feb. 11 there was a whoosh, the Turner house trembled, the sun was blotted out and a strange chill seeped inside. Garrett and his brother . . . forgot about hunting as their mother started them bringing in wood in a wind that had needles in it." Communication in 1899 being what it was, Dallas residents—and, in fact, people everywhere—didn't know much about the blizzard until it struck, but when it did, they quickly realized its viciousness. The precipitation that had been predicted for Friday the tenth arrived, but it came in frozen, blown by a cutting wind. By nightfall three inches of snow covered the icy ground. Homeless and jobless people sought warmth anywhere they could find it: in all-night saloons and in the courthouse, where they had packed into courtrooms during previously chilly days. A Dallas reporter described the sights and smells, especially a noticeable paucity of bathing manifested by a smell like that of "a fish fertilizer factory."

On Saturday the eleventh, the storm worsened, ruining a business day that merchants had expected to be bustling; in those years rural folk came to town on Saturdays to buy necessities for the following week. A thirty-five-mile-an-hour wind flung sleet along streets and sidewalks and against buildings, leaving an icy glaze. Most stores closed; only saloons stayed open, and they dispensed many a Tom and Jerry.

By four o'clock that afternoon thermometers had fallen to 20 degrees, and three hours later at least one instrument in the city recorded zero. The mayor ordered city hall doors left open and the interior heated for the homeless. The courthouse remained open as a refuge, while thermometers maintained a constant plunge. At 5 a.m. some of them registered 5 degrees below zero, al-

COLD

Snow is rarely seen in South Texas, but this snowball fight took place in Corpus Christi on Mesquite Street. 1897.

though they were unofficial; at seven o'clock, 10 below. Firemen answering a call from 110 Pecan Street were forced to build a fire around a hydrant to get water. Meanwhile, that house and two others nearby burned completely. At Fort Worth, frozen water tanks and frozen engines halted several freight and passenger trains. The blizzard moved on to grip the entire state. Men, women, and children caught outside froze to death unless they were quick to find shelter. At Mount Pleasant, a farmer named W. F. Jones set out on foot for town but died on the way. A McKinney doctor walked out to his stable to hitch up his horse for a trip and found the animal dead. In Waco a reporter for the *Weekly Tribune* later reported, plucking hundreds of heartstrings, "It was terrible for [people] from the abodes of poverty and destitution, and screams and wails of suffering went up." More fortunate citizens contributed wood, coal, medicine, and bedcovering to the poor.

Cattle as far south as Houston drifted with the storm until fences stopped them; two days later some of the animals, frozen dead in their tracks, toppled over when men sought to move them.

On Sunday, February 12, skaters reveled on the frozen San Antonio River and West End (later Woodlawn) Lake, but how the skaters got their equipment and the ability to use it poses a mystery to me, because residents of that normally sunny city have never been known as aficionados of winter sports.

South of San Antonio snow covered the ground to a depth of six or seven inches all the way to the Rio Grande and even deep into Mexico and beyond. Alice, Texas, reported six inches of snow, Laredo and Eagle Pass the same. Deep drifts near Austin derailed a Southern Pacific locomotive. Bays at Galveston (with the exception of the main ship channel), Matagorda, and Corpus Christi froze, and farmers in the Rio Grande Valley lost all their truck and fruit crops in a temperature that plummeted to 12 degrees.

Winter was indeed a bad guest that year, but cold can grip Texas without inflicting the extreme sadness and loss of 1899. Often it arrives in the form of that "blue norther"—and the term *norther* seems to be as distinctive to Texas (although other parts of the world have their own versions of "northers") as *santa ana* (hot wind) is to southern California, *mistral* is to France, and *nor'easter* is to the eastern seaboard.

Northers have been roaring out of the Arctic across the Great Plains and into Texas, carrying deep winter on their icy wings, at least as far back as recorded history tells us. A genuine *blue* norther can't be mistaken for anything else; or at least such a storm was rarely mistaken for any other by old-time Texans, even though they had no television or radio weathercasts available. Those people soon learned the dangers of being caught outdoors by one of those monsters, and they headed for home, or some other haven nearby, when they sighted an approaching threat: a long wall of clouds stretching across the northern horizon, looming ever higher, ever bluer, and ever darker in the winter sky. Whatever was left of broad daylight began to fade, and the wind would gust with increasing chill and violence.

These fast-moving northers don't just happen, of course: they have to be triggered. Low pressure in a central or southern region of the United States can do it by allowing cold and heavy air from the high polar area to surge down along a route east of the Rocky Mountains, usually across the Great Plains and into the Texas Panhandle. The fastest-moving northers can arrive from Canada in twenty-four hours. These are the winter storms that gave birth to those stories about the strange things that happen during sudden temperature drops. Usually, however, northers take up to three days for the trip.

Not all of them come packed with ice and snow. If Texas has had a westerly wind before the norther arrives, it has been said, the norther will often arrive dry. But a prior southeasterly wind, or moisture in the norther itself, or the effects of upper-air currents, or combinations of these things, can bring chilling rain, sleet, or snow—or a combination of all three.

Sometimes northers don't stop in Texas but blow into the warmer latitudes farther south: down the east coast of Mexico to the Bay of Campeche, into Yucatán, and occasionally, as happened in the 1899 blizzard, even farther. Central Americans call this a *norte,* I hear.

Ship captains in the Bay of Tehuantepec and several hundred miles farther into the Pacific Ocean can have more colorful and pertinent names for the storms, depending on severity. Some northers build up in the Bay of Campeche, roll westward through a natural outlet in the Sierra Madres of Mexico, and assault shipping in the Bay of Tehuantepec or in the Pacific.

Old-time Texans could describe peculiarities of northers, or thought they could, and considering their familiarity with the things they probably knew whereof they spoke. A dry norther, which they dreaded almost as much as one that brought sleet and snow, presaged its arrival with a display of those low, dark-blue clouds extending across the northern horizon, described in regard to a "blue norther." That was the name, in fact, that many old-timers reserved for a dry storm. Some of them declared they could smell a norther coming; it gave off the odor of burning hay or charcoal. Others thought a truer indication of such a weather change could be found in the unusual actions of birds or animals. Whatever the truth, much of Texas has had its share of sleet, snow, and ice every winter, and the frozen precipitation can make living uncomfortable for birds, animals, and human beings alike.

In December 1878, a heavy snowfall blanketed the Hale County area, north of Lubbock. One member of a beleaguered surveying expedition, O. W. Williams, later remembered the discomfort endured by the group's animals: "Our horses, staked out for the night, were gaunt and shivering. Our mules, tied to the wagon, [ate] the wagon box, part of the sideboards, and as much of our tent ropes as they could reach." (This was quoted by Wayne Gard in "Texas Weather," *Texas Parade,* November 1956.)

Eight years later, following more deadly winters, another very heavy snowfall struck the Panhandle. Cattle drifted southward against a vast and expensive fence built by ranchers to protect their ranges from such an event, and thousands of cattle died. A ranch manager, John Hollicott (of the LX), later skinned 250 cattle to the mile along thirty miles of the fence. In Hansford County, Texas, just south of the Oklahoma Panhandle, another man made a gruesome discovery: a settler's wagon with the team lying dead in the harness, and, inside the wagon, the frozen bodies of a man, woman, and three children.

Even with contemporary living improvements, a severe winter storm can bring anguish to Texans, especially to those who must go out in the weather, because many Texans can't cope with blizzard conditions as can residents in colder climes. Road employees have left warm beds at 3 a.m., bundled up, and joined crews that were tossing salt or sand on bridges and streets, only to see a high wind blow the stuff away almost as soon as it was thrown—as was reported in an Associated Press story out of Dallas on January 10, 1962. The day before, the story said, a major automobile parts dealer had sold twenty thousand sets of tire chains and "could have sold fifteen thousand more" if they had been available. His entire stock of antifreeze, six thousand gallons, had gone early in the day, but potential customers kept coming in.

Such cold weather can leave another bitter legacy with which automobile drivers must contend for weeks afterward: rough, broken streets and highways, damaged by extreme cold followed by quick thaws.

Among the many people who do not consider icy scenes beautiful are telephone repairmen. More than half a century ago, in 1924, when telephones in Texas numbered around one-twelfth of what they do today, an ice storm created this havoc: thousands of miles of ice-covered wires strewn over the ground, thousands of poles broken, and thousands of breaks in toll lines around San Antonio, Houston, and Beaumont. (Oddly, the worst destruction was limited to *South* Texas; it bypassed much of Northeast and Northwest Texas.) Repairmen bundled up and went to work in rural areas where animals had fled open fields for the warmth of barns. An old copy of *Southwest Telephone News* (February 1925) gave an indication of work accomplished throughout the area:

"The ice storm tore down 34,000 poles in Missouri [leaving St. Louis completely isolated], Oklahoma, and Texas. If we assume that the average number of cross-arms carried by each pole was three, or thirty wires to the span, . . . the total mileage of open wire put out of service was more than 25,000." So those repairmen of long ago faced the necessity of fixing wire long enough to encircle the earth at its equator.

The *Telephone News* added this observation: "Trees, fences, houses . . . [were] enameled with . . . glistening sheets. Dazzling diamonds of ice turned every shrub into a giant outburst. Glittering crystals of ice weighed down trees, transforming them into brilliant candelabra. Each blade of grass, as . . . sunlight [later] touched it, was a miniature radiant rainbow. . . . But . . . to telephone men who viewed it . . . it was . . . white hell."

During the previous winter season of 1923/24, Romero, in the Panhandle county of Hartley, recorded a total snowfall of sixty-five inches, highest ever in Texas.

Broad-ranging northers can make winters especially difficult in Houston and other Gulf Coast areas, where talk of weather usually concerns heat and humidity to such an extent that British diplomats get hardship benefits for being posted there. Residents might get used to the sultriness, particularly now that air-conditioning is so widespread, but the city is still unprepared for winter and probably will remain that way.

In December 1983, when Houston thermometers plunged into the midteens, an estimated fifty thousand water pipes burst in residences throughout the city. From three to five thousand homes had no water for several days thereafter because of a shortage of plumbing supplies. Largely re-

COLD

Effects of an ice storm in Waco. January 1, 1979.

sponsible for this domestic disaster in the first place was the construction of many frame homes, which were built a few feet above ground level to allow circulation of air. Water pipes were thus more vulnerable.

A long-time Houston resident I know looks toward winter with a bad attitude, and I think that's shortsighted. I love a little winter, since I am not a road-sander or salter or telephone repairman and do not live in Buffalo, New York. There is no grander weather than a gentle snowfall. On the other hand, I can realize why many residents of woeful-winter states like Minnesota, Wisconsin, Michigan, and elsewhere like to migrate for the cold season to the Rio Grande Valley, which usually remains reasonably comfortable in the winter, and, to me, exceedingly warm the rest of the year.

"It's estimated that the growing season in the Valley is 350 days yearly, with the other 15 days taken up in harvesting and turning the rich earth for the next crop," my friend Judyth Rigler, author, book reviewer, and Valley resident has written me regarding weather there.

> If you drive Highway 83 connecting the little towns and larger cities dotting the Valley—McAllen, Pharr, San Juan, Alamo, Donna, Weslaco, La Feria, Harlingen, Los Fresnos, Brownsville—you'll see the deep green tops of carrots waving at you one month, the bluish-purple of cabbage the next, the brilliant jade of green peppers only a few weeks later. Aloe vera fields abound in one section, testimony to the burgeoning industry they have birthed. The burning of the sugar cane fields is a dramatic sight—cane is harvested in this way, with the leaves of the stalks burned off at maturity—and if you're driving by, you'll have to stop and marvel. Those lucky enough to have a swimming pool at home don't consider themselves much blessed at this time, though; the tiny fingers of ashes from the burned cane drift through the air and downward like scorched snowflakes, coating everything with a fine black layer, including the thousands of turquoise pools that dot the area.
>
> The time of year in the Valley can be judged by the effluvia at the side of the interstate or the backroads. Jostled around in heavily loaded trucks headed for somewhere else, some of the produce slips out and settles on the edges of the roadway—like a certain percentage of the human population, rejected by fate. Cotton clinging to the roadway forms a false picture of winter Valleyites never see; busted (far more descriptive than burst) watermelons make a colorful, if somewhat macabre, scene; and onions—onions are everywhere, their aroma permeating the air, particularly on those rare occasions when it's damp.
>
> Some folks stop to take advantage of interstate gleanings after the field has been picked, carrying their treasures home to eat. A woman remembers growing up here during the Depression. They had little money, but the children of her family, walking home from school along the railroad tracks and packing sheds, would holler at the workers boxing up produce until the men threw vegetables their way. Mother would have a pot of water boiling when they reached home. With or without a small chunk of meat or a bone for flavor, dinner would be *caldo del rey*—soup of the king.

So the mountainless Valley, really an alluvial plain largely unfertile but for the miracle of irrigation, becomes a horn of plenty for the tables of families all over the nation. And that brings us to the second most visible offshoot of Valley weather, the tourists.

Called Winter Texans, Snowbirds, or, by the Hispanics, *Viejos* (Old Ones) or *Cabezas*

Blancas (White Heads), these retirees from the North and Midwest swell the Valley population by at least 150,000 yearly. The "Snowbird" title is perhaps the most descriptive; like migrating birds, they flock to South Texas to take advantage of the sun, the higher temperatures, the slower pace, the nearness of Mexico, the abundance of colorful tropical plants and flowers, and the availability of fresh, inexpensive food. Most—farmers and their wives from Michigan, Indiana, Ohio, Iowa, Minnesota, Wisconsin, and Canada—stay from November through April, when their own farms lie dormant or can be tended by their offspring back home. As common a sight as the fields of vegetables or fruits, the trailer parks stretch as far as the eye can see, with names chosen to attract the visitors from the North: Paradise Park, Border Breeze, El Valle del Sol, Tropic Star, Citrus Grove, Gulf Breeze, Los Amigos, Oleander Acres. When the citrus is abundant, many RV parks—former groves—offer a bonus: residents may pick the oranges, grapefruit, or tangerines from "their" tree—the one closest to their trailer. Palm-lined highways become clogged with couples in recreational vehicles, enjoying a busman's holiday as they drive along observing farming, South Texas style. Many Snowbirds used to go to Florida but now come here, where they say the climate is more favorable—less rain and more predictable weather—and the people are more hospitable, with the hand generally extended for a handshake rather than a tip. Valley weather reports on TV and radio compare temperatures "back home" with those here, and often there will be a 60- or 70-degree difference. Jackets are rarely needed; heavy winter coats, never. Still, at the slightest hint of cool, the Winter Texans sport their knitted caps and windbreakers with identifying tags for trailer parks and home cities and states.

By the time a change in winter weather reaches the Valley, there is usually very little left of its force. Visitors lingering over their third cup of coffee at an early-morning cafe marvel at the snowfall amounts and low temperatures back home, as opposed to the much higher temperatures here. Then they head out, shirt-sleeved, for a game of shuffleboard. Meanwhile, the same trailing cold front that brought twelve inches of snow to Ottumwa roars into the Panhandle, through North Texas, and into Central Texas. Around San Antonio it usually loses the last of its punch.

Occasionally in late summer or early fall, another form of low-pressure system kicks off a tropical storm, but the Winter Texans don't stay around for hurricane season. It's too hot for them here, and they must be home to plant and tend their crops. As the *Viejos* begin their slow trek northward after Easter, admiring the bluebonnets and Indian paints covering the roadsides, the Valley is moving into its six- to seven-month-long period of scorching, sunny days and warm nights, a sameness some find reassuring, others maddening.

Apparently I love *some* snow more than those Winter Texans do, but in my current location of Central Texas I can usually do little more than read about it. Through reading I have learned, for instance, that fresh snow reflects up to 90 percent of the sun's energy. A warming wind would thus have greater influence in melting the stuff than bright sunshine does.

I have learned other things through reading, but certainly I'll never know as much about snow as some physicists do at the National Center for Atmospheric Research in Boulder, Colorado. Recently they somehow determined that a cubic foot of snow, whether it fell in Texas or elsewhere, contains eighteen million crystals. (A cubic foot, in case you don't know, consists of one square foot and one foot in height.)

Who can call those scientists wrong? Other scientists. "Vincent J. Schaefer, dean of United States meteorologists, calculates that it takes more than one million crystals to blanket a two-square-foot area with snow ten inches deep," Ruth Kirk wrote in *Snow* (New York: Morrow, 1978), and other estimates vary, too.

Those same NCAR scientists who somehow counted all those crystals in a cubic foot can tell you at what temperature each crystal was formed, as categorized by its appearance: six-sided plate, star, column, needle, or bullet. ("It is known now," Kirk wrote, "that six sides are basically inherent in the atomic structure of snow crystals, although some are three-sided or five-sided.")

And those same NCAR scientists contend that it is quite possible that no two snow crystals have ever been alike. Kirk and some other researchers disagree. Considering the myriads upon myriads of snow crystals that fall each year throughout the world, and that have fallen since winter first began, who can be imperious enough to assert that no two flakes have ever been exactly alike?

This argument does not deter the NCAR people. They say it could be possible, since each snow crystal in any given pile has trillions of water molecules that could have frozen in different ways. By multiplying the number of water molecules in a crystal by the many ways they can freeze, they say, it becomes possible that no two crystals have ever been exact duplicates. My research calling them wrong hasn't been completed yet and might never be, but I *do* have a suggestion regarding their work with all those impossible figures. It should go into solving the problem of the national debt.

Easier to comprehend is some information I have just now read about snowflakes. They are not necessarily snow *crystals*. The authority is Kirk: "A flake is an assemblage of individual crystals, both whole and broken, joined together in falling."

So we are both learning something about snow—you and I. I never knew that flakes and crystals are usually different. I *do* know that when they all melt, water is formed, and that brings us to the next chapter.

11. WET

The man of some education who once suggested a subway system as a solution to traffic problems in Houston obviously was not a resident of the place. He must have been suffering from the delusion that Texas is dry all over.

In recent years Houston has become one of the most flood-prone areas in the United States. Give the city a thirty-minute drenching and many vital streets, especially in southwest Houston, will become surging streams, stalling cars and making driving impossible for a while. Let it rain for hours, and bayous begin creeping toward adjacent residential areas.

Maybe the man meant to say sub*marines,* not sub*ways.*

Houstonians are rarely as happy to get rain as, say, San Angeloans, whose newspaper, the *Standard-Times,* used to print a red silhouette of a rooster on page one to celebrate yesterday's downpour. Houston has not always been as flood-prone as it is now, though accounts left behind by early visitors make it clear that the coastal rains always did transform dirt streets into squash and made a mess of horse and mule and ambulatory travel.

What has happened to Houston (and many other cities) is that so much concrete and asphalt have been laid down that water has fewer places in which to soak into the soil. Consequently, giant downpours of almost any length of time can leave the city awash for varying periods.

All this compounds the original problem: a low-lying, flat terrain. The highest elevation in Houston is around thirty feet above sea level, and parts of the city and surrounding Harris County are losing even some of that elevation as water is pumped from beneath the surface. Sections of the city lost as much as a foot of elevation between 1973 and 1981 (*Time,* November 30,

1981). The flood-control district maintains six thousand miles of drainage ditches and bayous, but adequate runoff is hard to come by, particularly when high tides on the coast a few miles distant act as barriers.

Some personal experience helps me to appreciate and understand the Houston problem, although my trouble found an easy way out—and I actually had nothing to do with the solution. The solution, in fact, arrived entirely unexpectedly.

A year or so after we moved into a new home in 1968, the first heavy rainfall bared an unforeseen danger. One night, during a giant summer thunderstorm when rain fell so thick and fast we couldn't see our backyard peach trees even with the porch light on, water began to rise alarmingly on the concrete floor of the patio that adjoined the house. Rainfall earlier had resulted in a slightly flooded patio, but this time the level began edging toward the sliding glass door leading to the kitchen, which was built on a slab (along with the rest of the house) elevated only four or five inches against such a threat as now presented itself.

This time the rain fell much faster than the excess water could drain off through a wire gate that served as an exit from the tile-fenced backyard. The depth on the patio increased, and still it rose—until the first small puddle showed on the kitchen floor, then a broadening puddle, then a stream. Sheets, towels, and heavy clothing dumped on the floor soaked up some of the overflow, but these things had to be picked up and squeezed in the kitchen sink, then thrown down again.

The thunderstorm rumbled elsewhere, and the water finally receded. Mops and brooms took care of the mess that remained.

Thereafter, instead of welcoming rain, I feared the darkest and most ominous clouds. I examined the backyard to see how landscaping might prevent a recurrence. Twice more in the following year or so heavy rains brought a frustrating repetition of the overflow, but at least it lessened the necessity for mopping the kitchen floor very often. Then, one late-winter day, our daughter Mary, whose avocation is gardening and keeping animals and fowl of various kind on her rural acreage, brought over a gasoline-powered tiller and plowed much of our backyard for gardening. The soil, hard yet covered with a tough mat of grass and weeds, disintegrated into clumps of earth only after a tussle.

Since then the space has been plowed twice a year, and the overflows have ceased. We can only presume that water now soaks more easily into the garden area and doesn't pile up against our backyard gate and kitchen door. Either that, or it really doesn't rain the way it used to, as old-timers are fond of saying.

But throughout most of Texas, it *does* rain the way it used to—maybe more so. Any Houston resident would surely agree, even without the asphalt and concrete that has been laid down. And weather records verify it, despite all my talk about drought. In 1982 the statewide mean for rainfall was 26.2 inches; in 1979, 31.4 inches. (The 1892–1979 mean, according to records in *One Hundred Years of Texas Weather, 1880–1979*, was 27.1 inches.)

Difficulties arise, however, when we try to estimate any sort of *mean* rainfall for the entire state. Clarksville, in Northeast Texas near the Oklahoma line, has recorded 109.4 inches of rain in one year, 1873. Three sites in Brazoria County, on the Gulf Coast south of Houston, have recorded more than 100

House going over the Austin Dam during the Colorado River flood of June 15, 1935. The reconstructed dam was later named for Tom Miller, the mayor of Austin from 1933–49 and 1955–61.

inches (Freeport, 106.4, 1979; Alvin, 102.6, 1979; Angleton, 100.2, 1973). But some locations in West Texas have withered in the sun for a whole year, dampened only by birdbath amounts. In 1956 Presidio recorded 1.6 inches; in 1910, Ysleta (near El Paso), 0.7 inch. Such is the diversity of Texas weather.

Parts of the state can be *very* dry some years and *very* wet other years. An impression I have is that the area around Lampasas offers one such example, but I haven't looked up rainfall records there to verify this. From long ago, in the mid-1930s, when my parents and I traveled a gravel highway through the town during summer visits here and there, I remember Lampasas for its brown grass, dust-coated trees, searing heat, and soft drinks that I drank, at a nickel a bottle, as long as pocket change held up.

A renowned Lampasas native, newspaperman Stanley Walker, hated those dry years after he had retired and moved back from New York, where his journalistic talents had propelled him to national fame as city editor of the *New York Herald Tribune.* At this moment I am looking at a clipping from the *Herald Tribune* of August 27, 1956—a story by Stanley Walker about the effect of a seven-year drought on Lampasas and on his retirement ranch there. "One must search the ancient, unverifiable Indian legends for anything comparable to this. . . . For many stoical though usually hopeful citizens it is the final turn of the screw. They are finished. It is as if they had been finally ground down by some burning force that moves as inexorably as a glacier." Although there was some grass, Walker wrote, it was short and yellow "and quite useless." Cattle and sheep living on it alone soon died, or became wasted, paralytic, and idiotic in their movements.

Less than one year later, Stanley Walker was reporting for his old newspaper on a flash flood that swept through Lampasas Sunday night, May 12, 1957, behind a ten-foot wall of water and inundated much of the town of five thousand. It struck "with the force of a tidal wave." The big villain was Sulphur Creek, which cut through a levee following a six-inch downpour. The wall of water then swept diagonally through town.

A dramatic but eventually humorous reminiscence of that terrible night has been given to me by Jane McMillin, then a second-grader, the daughter of a Lampasas doctor, Rush McMillin. Guests of her family at a dinner that Sunday evening included Dr. Allen Spence, a young intern at Rollins-Brook Hospital, where her father practiced.

Thunder and lightning became scary to her soon after the storm moved in, and rain that had begun falling around dark became a deluge. Lights in the house went off, and soon after that Dr. McMillin received a telephone call from the hospital. The backup generator there was giving trouble; nurses were taking turns hand-pumping an iron lung to keep a patient alive. McMillin and Spence left for the hospital and arrived soon afterward. Mrs. McMillin and a neighbor, Ruth Rawlings, ventured into the weather to comb the neighborhood for candles and lanterns—anything to provide emergency lighting for patients and doctors.

Later Jane McMillin heard about what happened during the rest of that night and reconstructed the story. Someone told Dr. McMillin that a colleague, Dr. W. M. Brook (popularly known as "Doc"), had ridden in an automobile with Mr. and Mrs. Ryan

Howard and their son, Donald, to watch the rising water and was missing after the flood had caught them. McMillin left Spence at the hospital and left with the Reverend Ernest Miller of Central Christian Church and Mrs. McMillin to search for the doctor. They used Spence's pickup truck. McMillin and Miller tied a long rope around their waists and crept through the dark water, calling out, while Mrs. McMillin, in the pickup, fed out the lifeline.

Meanwhile, all over the blacked-out town, water swirled, carrying away residences, ruining merchants' stocks, drowning five persons, injuring others. In a seventy-five-year-old stone hotel water rose nearly to the second floor. Around the courthouse square floodwaters tossed automobiles "like chips" (Stanley Walker wrote). Lumber dealer B. K. May, fearing trouble, had gone to his office and was there when his entire business was swept three blocks by the flood. He saved himself by swimming to a telephone pole that somehow stayed erect.

Dr. McMillin and his wife and the Reverend Mr. Miller gave up their search. McMillin returned to the hospital, where he helped treat victims suffering from exhaustion and exposure. One of them was Mrs. Ryan Howard, who (with her husband and son, who also survived) had been with Dr. Brook when their automobile was flooded.

The ending to Jane McMillin's story is a happy one—even funny—but no doubt the humor wasn't fully appreciated until long afterward. Dr. Brook survived the flood; for hours he had clung to a tree near a residence, yelling for help. But the woman of the house had mistaken his deep, gravelly shouts for the croaking of a bullfrog.

History records many other disastrous rainstorms. One of them was a 1921 downpour apparently centered two miles north of Thrall, in Williamson County, thirty-five miles northeast of Austin. It is still mentioned in weather records. The rainstorm, which flooded areas around San Antonio northeastward into Austin and Taylor, grew out of remnants of a weak September hurricane that had crept in from Mexico and South Texas.

My friend, John Banta, who retired several years ago as Central Texas editor of the *Waco Tribune Herald,* once visited Thrall and talked with an eyewitness of that storm, a man named F. A. Nolte. Nolte, then a twenty-six-year-old farmer, watched the turbulence roll in early one morning without much initial excitement. He stood on his front porch. "It started . . . misting," he said. "Then it got heavier and heavier. The wind didn't blow much. There was a right smart . . . lightning and thunder, but I've seen more lightning than there was in that cloud."

As the new day grew darker, rain began falling faster. "You couldn't see anywhere, it got so hard." Midmorning gave way to noon, then midafternoon passed, and still the rain fell. The surrounding area, during the rare times it became visible at all, appeared to be a vast, stormy lake. Every low area had been inundated. Shallow trails made by cattle became raging streams capable of drowning people. Water ran everywhere, but Nolte's soaked house stood on a rise.

Visiting at the time was an uncle from Austin County, near the Gulf Coast just west of Houston. The man joined Nolte on the porch to watch the incredible rain. No coastal downpour he had ever seen compared with it. Nolte remembered that his uncle, Harmon Nolte, paced the porch much of the day, peering at the rain and exclaiming from time to time, "My God! When is it going to quit?"

It didn't quit for hours and hours thereafter, but climatologist John F. Griffiths and Greg Ainsworth, who also researched that calamity and described it briefly in *One Hundred Years of Texas Weather,* said the deluge came in "two bursts of great intensity" in the area around Thrall—"one about 0630 on the 9th, the other at 0400 on the 10th." Measurements accepted as official showed 19.6 inches in twelve hours and 23.1 inches in twenty-four hours, but unofficial totals went as high as 32 inches in twelve hours and 38.2 inches in twenty-four hours, "occurring some two miles north of Thrall," the authors continued. This was the approximate site of Nolte's farmhouse.

(Someone's measurement of 36.4 inches during an eighteen-hour period was sufficiently official to get into record books, and it is still there; I have seen many references to it throughout the years.)

Nolte himself had no official gauge, but he had his own awesome indicator of how much rain had fallen. Before the storm struck, he happened to have placed an uncovered fifty-five-gallon barrel, one-third full of water, in an open area near his house. When the rain finally stopped, he saw that the barrel was running over. He heard neighbors nearby say that the deluge had filled several of their previously empty fifty-five-gallon barrels.

Nolte remembered very well the damage done to the land. Huge gullies had been cut through fields and pastures. Shallow swales had become yawning ravines "eight to ten feet deep." Many fields became cultivable again only after years of work. Worse, 215 individuals lost their lives. Property damage amounted to $19 million, but that really tells nothing in regard to today's dollar. Homes, crops, bridges, roads, and railroad tracks disappeared. Oddly, however, October, November, and December proved to be much drier than usual. And even the September storm affected only a comparatively small part of Texas. The San Gabriel and Little rivers flooded, sending the Brazos River out of banks southward. Elsewhere, high water flooded parts of San Antonio, San Marcos, Austin, Hearne, Georgetown, and Belton; but Waco, not far to the north, received only 0.98 inches.

More than half a century later another Texas town—Alvin, between Galveston and Houston—would replace Thrall as the unofficial (but Weather Service–investigated) site of the heaviest rainfall: forty-three inches in twenty-four hours, July 25/26, 1979, also resulting from a hurricane—Claudette.

Farther south down the Gulf Coast, one 1984 rain of between fifteen and twenty inches in twenty-four hours drenched Robstown, just west of Corpus Christi; eyewitnesses claimed that the sudden deluge sent the Nueces River in that vicinity running backwards.

On the other hand, Texas has those areas where annual rainfall drops below ten inches and the land becomes a desert. Ten to twenty inches annually can nurture enough grass for grazing livestock; farming usually requires more than twenty inches.

Water is essential in many other ways. Adult humans usually require between two and three quarts daily, including water from food eaten. (Our bodies are made up of about two-thirds water, approximately

Looking north from South Congress Avenue in Austin, during the Colorado River flood of September 28, 1936.

speaking, and that description can also be applied, loosely, to the earth's surface.) In a typical lifetime, if such wording can be used, this means an intake of something like sixteen thousand gallons of water. Add the requirements of farm irrigation, industry, and other activities to this, and it is easy to see why some thoughtful Texans foresee water as the state's number one limitation on growth as we near the twenty-first century.

My own affinity for plentiful water has already been expressed more than once. I imagine many other Texans share that feeling, and I *know* one who does. After I had begun writing this book, I wondered about the weather preferences of our five children and discoverd another rain worshipper in our eldest daugher, Carol. She wrote several paragraphs on the subject with absolutely no knowledge of what I had already written for the book, and with no help from me; and the similarity of her and my weather preference is remarkable (especially when you compare what she has written below to what I wrote later, in a section you will find in the following chapter, about a 1953 Waco tornado).

> My brother, sisters, and I became weather watchers early in our lives. We learned from the childhood summers spent with our grandparents [*my* father and mother, of course] on their Central Texas ranch. Every day we would drive to the ranch from Temple, where they lived, to check on the cattle and keep our weather vigil. The grass would usually be brown and dead, the ground riddled with deep cracks, the once-flowing creek dried up into small pockets of stagnating water, and the blue sky like a lid keeping the heat in.
>
> Faced with thirsty stock and a water shortage, there were times my grandfather would manually pump enough water to fill a large cement trough with water when the breeze was not sufficient to drive the windmill. The water would come out cloudy and with a strong sulphurous smell, but it would nevertheless bring the cattle in from the far reaches of the pasture.
>
> Other times we would walk to the dry gravel creek bed, we kids taking turns carrying the shovel, and "dig down to the springs" so that enough muddy water would bubble up to slake the thirst of a few cows.
>
> I remember seeing smoke curling ominously against that incessantly blue sky, warning of a distant grass fire. We would watch with dread, knowing that if it spread to the ranch we would be mostly helpless to put it out.
>
> Most of all I remember weather watching. I can still see my grandfather, tanned face and silver-gray eyes, scanning the westward horizon. Some days big fluffy clouds would roll in with a promise of rain that was rarely delivered.
>
> But my favorite weather watching was done at night. Back in town, just before going to bed, we would follow my grandfather outside to peer at the sky again. Often the stars would be glowing against a black velvet background, but that was not what we wanted to see. An absence of stars would have meant clouds, possibly bringing rain. Sometimes we would see heat lightning off in the distance and would wish anew for some thunderstorm to wet that dusty and dry ranch fifteen miles to the north.
>
> Eventually we would be blessed with rain. Sometimes the showers were so slight that pockmarked dust would be the only evidence of it. But there were also the joyful days when the rain would pelt us as we rushed to the car to get out before the dirt road turned into a quagmire. Once on the pavement we would watch, soggily but

gratefully, as the rain washed clean our worries.

I am today a lover of rainy days. I attribute this quirk to my early summers. Even years of living in Southeast Texas, where too much rain can present problems, have failed to change that. Friends who do not share my dry summer background are amazed at the equanimity with which I accept rained-out canoeing, hiking, and camping trips. It is not that I don't enjoy canoeing, hiking, and camping—they are my passion—but so is rainy weather.

Just the other day, though, after a week of steady thunderstorms, I left a postgraduate class in the Houston Medical Center and was walking a mile to my automobile. I splashed through puddles and slid through mud and remembered the summers of my childhood. I was halfway to the parking lot when the bottom dropped out of yet another thunderstorm, and as I ran, book bag pounding my back with every step, water streaming through my hair and with glasses fogging up, I did—just for a minute—wish for an umbrella!

Reading that is like holding up a mirror to my own weather experience and desire. *My* most urgent yearning for rain in Central Texas, however, took place while I was fifteen hundred miles away, in California, where I attended junior high and high schools, 1937–42. I had spent part of the summer of 1937 on my own grandfather's farm, the one I have written about (some twenty miles south of the ranch that my father eventually bought). To while away the days awaiting departure for California, I hit upon a game of sorts that would appeal only to a thoughtless lad who hadn't even reached teenaged maturity.

Shells for my uncle's .22 rifle could be bought very cheaply in those days, and I had enough money to afford a box or two. At first I aimed at birds happily chirping on tree limbs, but fortunately (as I consider it now) I didn't hit any of them before realizing, even with my then-limited thinker, that such high shots could be dangerous to unseen human beings and cattle standing somewhere in the vicinity but out of sight.

Looking for other targets, I saw, in the middle of my grandfather's stock tank, poking above the surface, the head of a turtle. My reading about that time had centered on World War I U-boats, and I aimed the .22 rifle carefully but venomously. There, in the middle of that stock tank, was a German U-boat waiting to inflict great destruction and death. But this one, I vowed, would never again sink one of our ships.

The trigger-pull brought a sharp "Crack!" that I immediately translated into the burst of a three-inch naval gun, or maybe the roar of a five-incher, or the rumble of a depth charge. The periscope vanished immediately. No wreckage appeared, but I recognized a direct hit when I saw one.

Two or three other U-boats appeared, to their everlasting sorrow, but then the pack apparently dispersed. No more periscopes showed themselves.

All things, however, come to one who waits long enough, and for a youth with short-fused patience, sometimes it doesn't take long at all. After an interval a battleship hove into view. There it was: across the stock tank and off to the left—a brand-new washtub my grandfather had bought with a precious depression dollar or two (I learned later) for performing a variety of jobs like laundering clothes and watering animals and so on.

Actually the new battleship lay in dry dock, about three feet from the water. I walked over, christened and launched it for the German navy, returned to my true station on the opposite bank, and aimed salvo after salvo at the monster until it had taken on so much water that it sank in mid-ocean, about where the first periscope had disappeared.

The kaiser's navy lay at the bottom of my grandfather's stock tank, and only I knew it. That was my good fortune, as I soon discovered.

I returned to the house about the same time that my grandfather embarked on a chore that required the use of his washtub. After a while he asked around if anyone had seen the thing, which he thought had been left down by the stock tank. The only person he didn't ask was me, and I didn't volunteer anything. After an hour or so he began referring to his washtub the way he sometimes did to his mules, but that inanimate object proved to be just as stubborn as the animals. He never found it, and it was then that I overheard the details of its newness and cost.

Another dry spell had begun its squeeze on that part of Texas. Before we left for California, the stock tank had dropped several inches more, but without revealing its grim secret. I left the farm praying for rain. My grandfather only squinted at the sky in silence, as usual without profaning the weather, and he had even stopped wondering about his vanished washtub when we departed for the Golden West.

In California, we began getting those penny post cards from him with the scrawled messages. The first one announced a recent thunderstorm, which led me into overconfidence. Subsequent cards told of "no rain in two weeks" . . . "three weeks" . . . "a month," and so on. After that I waited daily for the arrival of extradition papers, but no further bad news came. Obviously enough water somehow remained to cover the sunken hulk, or else my grandfather had gone blind—which was an impossibility, I knew, because nothing ever happened to him.

Then came a scrawled card with the joyous news that a three-inch rain had soaked the farm, or maybe it was four inches or even more, and had left the stock tank full.

To this day that ranks as the sweetest rain that ever fell on Texas, and I wasn't around to feel, smell, or taste it.

12. STORMY

Ever since the first chroniclers began recording their eyewitness observations of events in Texas, the storminess that occasionally plagues inhabitants has been reported. Detailed portions of some accounts of rainstorms, snowstorms, and windstorms have already appeared in this book, but other types of storms—hurricanes and tornadoes and lightning—have caused many more deaths and much more property damage during shorter time periods.

Nobody living today can describe firsthand what were once two of the most talked-about hurricanes ever to strike the Texas coast—in 1875 and 1886, at the then-thriving port of Indianola (forty miles southeast of Victoria), which had been built three or four decades earlier facing eastward toward Matagorda Bay.

In 1875 the town counted a population of several thousand inhabitants. Vessels arrived at and departed from Indianola on voyages that linked the town, via Galveston and New Orleans and Havana, with exotic ports of trade all over the world.

Indianola had been built on low and sandy soil, but residents ignored the potential danger of this. Standing as flimsy protection between their townsite and the Gulf of Mexico lay only a strip of more sandy land: Matagorda Peninsula and Matagorda Island, broken by a pass (Cavallo) through which vessels entered into and exited from Matagorda Bay. Residents experienced some flooding every year because of high tides and winds, but damage always proved minor, and this gave them fatal confidence for surviving even serious hurricanes.

Anyone who knew the Gulf Coast and Gulf storms should have been able to foretell disaster, it seems, but the general prosperity and the diverse pleasures that Indianola afforded, plus the good luck with weather, closed most eyes to the possibility.

Besides, the site of Indianola had some two hundred years of history behind it for a supposedly solid foundation. In 1685 La Salle probably landed somewhere nearby, around Matagorda Bay and possibly at the subsequent site of Indianola. In the 1820s and 1830s some of Stephen F. Austin's colonists came ashore there. Then, in the midnineteenth century, when its name had changed from the original Powderhorn to Karlshaven (still later to Indianola), German immigrants landed; some settled, others went on to new homes farther inland. Morgan Lines vessels began using two long wharves extending into Matagorda Bay. In 1856 the United States War Department offloaded two shipments of camels to begin an experiment in desert transportation between southwestern military posts. Three years later rail service began. Indianola grew despite many nineteenth-century threats: yellow fever and cholera epidemics and Civil War coastal fighting. At its peak shortly before the 1875 storm, the port, which was also a county seat, had become the most important town on the Texas coast. South of it a bayou cut inland for half a mile or more, circling like a powderhorn toward a long lake, abundant with oysters, situated behind the town and southwestward from it. North of the lake, but merged with it, lay an area of marshland.

Most business houses, generally one- and two-story frame buildings of typical nineteenth-century frontier style, lined both sides of Main Street, which stretched on a north-south slant parallel to the bay. A few exceptions to the frame construction stood out. One was the Calhoun County Courthouse, made of a crumbly kind of brick in which crushed sea shells had been used. The courthouse faced the bay on Water Street, just eastward of and parallel to Main, along with some flimsier structures.

Most of the businesses were situated at the southern end of the stringlike town, near the wharves that poked eastward into Matagorda Bay. Before 1875 cargoes included ice cut from New England lakes and shipped, insulated as much as possible, from Boston. No other ice was then available.

West of Main Street (toward the lake and marshland) one or two abbreviated streets paralleled Main for short distances. Farther west, beyond the marshland, a depot and some machine shops had been built alongside the railroad. Connecting tracks linked it with the wharf area.

All this described, in a general way, the smug city of Indianola in mid-September 1875 when a severe hurricane howled into the area and rudely awakened inhabitants to the weakness of their sand-built town. Many years later a survivor, Mrs. Henry Saunders, recalled the terror in a May 16, 1926 interview in the *Houston Chronicle*.

Wind, rain, and tidal flooding increased steadily on Wednesday, September 15, but few persons became alarmed, she said, "as we had often known the water to rise waist high during the spring and autumn season." Despite the inclement weather, she left home to visit a friend who lived near the lake. That night she saw that a high tide would prevent her return.

On the following day the storm showed itself to be more than just another seasonal tide-tosser. Seas poured over Matagorda Peninsula into the bay, and the water swept across the low, now-unprotected beach along which stood Indianola. Water also roared up the bayou into the lake and sent an overflow sweeping into the marshland.

STORMY

This deadly tornado ripped through Wichita Falls at 6:15 p.m. on April 10, 1979, killing 44 persons, and leaving 5,000 homeless.

Mrs. Saunders paused to ask her interviewer, "Perhaps you have noticed . . . two cedar trees on the south side of the bayou [near the marshland]? In [a] home that stood there 65 people were saved. [It] was a two-story building and the family with their terror-stricken guests . . . fled to the upper floor as the water rose higher and higher in the house. They leaned from the upper windows and grabbed less fortunate friends who were [being] hurled along on pieces of driftwood. . . . Many people knocked their homes from the foundations to prevent the crash that inevitably crumbled the homes if they fell. Some bored holes in the floors so that they could not float away."

The 1875 hurricane ended the lives of anywhere from two hundred to seven hundred individuals around Indianola, according to contemporary estimates, and swept away three-fourths of the town. Only underground cisterns, hurricane-polluted, and oleander bushes remained to show many residents where their homes had stood. Storm-whipped floodwaters from the bay, not the high wind, caused most of the destruction. That particular hurricane didn't wipe out all optimism, however, and many survivors rebuilt. The town established on sand lived another eleven years, although after 1875 it lost some importance and bustle even beyond that caused by water and wind damage.

By August 1886 the bay had encroached toward Main Street, eliminating most of Water Street, though not the courthouse. Main still displayed the stringlike cluster of stores and houses that marked the central part of town. Only one wharf poked into Matagorda Bay to receive vessels, but their arrivals were less frequent now. The second disastrous hurricane blew in on Thursday, August 19, and added damage by fire to water and wind destruction.

Manning the Indianola weather office that evening was Isaac A. Reed of the United States Army Signal Corps. Reed could provide little help in battling a hurricane, other than sending readings to the central office in Washington, D.C., but he died trying to fulfill that duty. The assistant collector of customs at Indianola, T. D. Woodward, later told the story in an official report sent to Reed's commander in Washington.

Woodward and a government telegraph operator had been sleeping in the telegraph office, next door to the weather (or signal) office, the night of August 19/20, when Reed woke them about three o'clock with a request to transmit the most recent readings of a now-terrifying tropical storm. "The instrument would not work and we were unable to do so," Woodward said in his report, which has been reprinted in the *Indianola Scrap Book,* compiled by Leonard Joe McCown (Port Lavaca: Calhoun County Historical Survey Committee, 1974). "We tried off and on from then until 4 a.m. but without success. The cyclone increased with such rapidity that we were afraid to remain longer in the telegraph office so went over to the signal office and assisted Captain Reed in taking his last observation, which was done at 5 a.m. . We then went downstairs, and remained standing [with an Indianola doctor, Halsey Rosencrans] in front of the building about five minutes when [it] fell in." Woodward and his telegraph operator realized their peril in time to run for comparative safety elsewhere, but Reed and Rosencrans "were killed, drowned, or burned; we don't know which happened first, as the building caught fire from a [kerosene] lamp we [had used] to read the barometer." The fire spread and burned adjoining buildings.

"The wire attached to the anemometer broke at about 3:30 a.m. and the last register was 72 miles an hour. Captain Reed leaves a wife and two . . . children. . . . He perished literally at the post of duty, the last observation being taken about five minutes before he was killed."

The storm wind that combined with water and fire to destroy Indianola for the second and last time surely blew at a greater velocity than seventy-two miles an hour, but no one ever knew the actual speed. It was powerful enough to send a locomotive and tender rolling from a wharf into the bay. Survivors' accounts indicated without doubt that the 1886 storm struck much more viciously than the one in 1875. Since the population and importance of Indianola had decreased greatly, however, death and damage tolls were not as great, although every structure in town suffered damage and most were demolished. Some homes left largely intact were moved to Port Lavaca, repaired and refurbished for occupancy, and inhabited for years by former residents of Indianola or their descendants.

Today you can strain your eyes for signs of old Indianola, but there are few to be seen. I visited the site about 1966 and examined a disintegrating cemetery and remains of what was apparently a cistern. Offshore I could see a crumbled ruin that must have been what was left of the courthouse. If the town could be magically reinstated the way it was, I imagine much of it would lie in Matagorda Bay these days.

A year or so before my visit to Indianola, a friend, Arnold Rosenfeld, then a reporter for the *Houston Post,* looked over the scene, talked with some people who knew the area, and wrote "Indianola Remembered" for his newspaper. He talked with a local resident, Dan Zimmerman, then sixty-nine, who remembered when a section of the courthouse stood thirty feet above the water. But "the sea now breaks over what is left."

And he talked with a Houstonian, A. M. Cox, who owned a beach home nearby. Cox occasionally saw the locomotive lost in the 1886 storm, but not very often. "Sometimes when the tide's low you can see [it] come out of the water. It happened a couple of weeks ago."

The Indianola of the nineteenth century has vanished except for the dwindling reminders that sometimes force a backward glance. For the most part, only ghosts remain. But to any history-minded individual who stands near the sighing surf and recreates in his mind the old town that once stood there, there are ghosts aplenty, and they will never vanish.

Nor will the ghosts of 1900 Galveston.

Many other hurricanes have hit the Texas coast. On September 4, 1766, a storm destroyed a mission on Galveston Bay, and in September 1818 a hurricane sent salt water racing four feet deep across Galveston Island, to the sorrow of pirate Jean Lafitte and his friends. And so on.

Other hurricanes are sure to strike in the future. My best advice about them, after personally experiencing Hurricane Carla in 1961, is to board windows, take any other precautions you can think of, and get out, especially if you live near the water.

In recent summers National Weather Service meteorologists have been talking of an additional threat stemming from overconfidence on the part of residents of coastal

The ruins of the San Antonio Loan and Trust Building at Navarro and Commerce streets, following the stormy spring weather in May, 1900.

areas. Many natives have never experienced a truly vicious storm; many others have moved into those areas from other parts of the United States and don't realize the capabilities of a major hurricane.

Few cities have efficient, up-to-date evacuation plans, and the National Weather Service claims an ability to give only twelve hours or so advance warning of a landfall. This time period wouldn't allow complete evacuation of a city like Galveston even if residents began leaving at the very first minute of warning—which is a reaction too immediate to expect.

Those same meteorologists also warn of a danger not generally recognized: that wind and water usually cause less damage than does the "storm surge," a giant dose of water sometimes fifty miles long that is formed by winds of great force. Surges have been referred to as tidal waves, but their actions are different. They cause sudden rises that can flood coastal areas to heights sometimes as great as twenty-one feet above sea level. Atop these rises powerful five-foot waves compound the battering effect, although not many of these details are visible to the human eye at the time of occurrence because of windswept spray that renders the surface and the atmosphere indistinguishable. Few structures can survive such surges, director Neil Frank of the National Hurricane Center said in 1982, and he also remarked that no building code had been written then to take them into account.

It seems to me that in many cases the only safe building code is *not* to build, with the Texas Gulf Coast as low-lying as it is. Large sections of the cliché-famed rockbound Maine coast might not be inundated by a storm surge, but much Texas coastal land would, and does, disappear beneath waters whipped up by a hurricane.

It was such a storm that caused much of the damage at Galveston that disastrous Saturday, September 8, 1900. In 1956, when I was involved in researching *A Weekend in September,* I examined original documents in the Galveston Weather Bureau office left behind by Isaac Cline, the meterologist in charge in 1900. Cline wrote in his "daily journal," used for submitting official reports, that about 6:30 p.m. Galveston time (7:30 p.m. Washington, D.C., time, which always went into the reports) a storm wave sweeping ashore "in advance of the hurricane vortex" caused a sudden rise of four additional feet, and shortly afterward the entire city lay under water to a maximum depth of more than fifteen feet. This storm wave caused much of the damage eventually reported, Cline guessed, and eyewitness accounts bore him out.

Involved indirectly in that Galveston hurricane was my weather-observing grandfather. By Saturday afternoon storm clouds moving in from the Gulf had begun dumping rain on his 1900 farm, then situated a few miles north of the acreage where he later erected the frame barn and used it to record those observations. But this rain, which was probably welcomed at first, proved unfortunate. My grandmother, so I have heard many times, lay abed awaiting the arrival of a fifth child, and the family doctor was nowhere around. He was in Temple, some ten miles west, and no telephone was available to summon him.

A neighbor volunteered to saddle a horse, ride into town, and get the doctor. Ten miles he rode through an increasingly violent storm. Rain made the black Bell County soil a quagmire; wind battered him and even felled some small trees standing along the muddy road. The trip required two and a half hours.

The doctor saddled his own horse and rode into the storm as soon as the neighbor delivered his message. He arrived at the farm in time, but flooded creeks prevented his returning to Temple for two days.

I never saw my aunt who thus came into the world. She died at the age of twelve, during one of those epidemics that used to sweep across the land. But I've heard her name all my life. It was Susie Gale—not Gail, or Gayle—*Gale,* deliberately spelled that way to commemorate the storm during which she was given life while, two hundred miles southeast, thousands of Galvestonians were losing theirs.

Hurricanes have had another effect. Texas is getting smaller because of them. Geologists estimate that in "only" 18,000 years the Texas coastline has given up to the Gulf of Mexico from eighty to one hundred miles of land; one reason for this would be the tropical hurricanes.

In recent years erosion has picked up noticeably. People who were once children growing up near the Gulf have returned for visits to the homes of their youth and have been surprised to find that entire beaches where they once played are no longer to be seen. Sometimes homesites themselves have disappeared. Galveston Island reportedly loses from five to ten feet of shoreline every year. Majority opinion among authorities seems to have it that erosion cannot be contained—certainly not permanently. Attempts at containment have been made, and sometimes with temporary success. One unusual recent method is using old Christmas trees collected after the holiday season and staked down on beaches to prevent erosion of sand dunes.

Hurricanes and talking about them constitute a way of life along the Texas Gulf Coast, even if many new residents have never seen a truly vicious tropical storm.

"Our section of the country seems to have one more season every year than other sections—the hurricane season," said Lucile G. Lyda, now of Austin, who recalled for me the days when forecasts and communications were on the primitive side. She sent me a reminiscence she wrote for *Adult Services Advocate,* Austin, November 1984.

> I remember an unnamed storm in late August, 1933. Another couple and we, with our three-months-old babies, were in Matamoros, Mexico, on vacation. With no car radio and no translatable newspaper, it was some time before we realized the significance of the gale winds and pounding rain. We started back to Corpus Christi amid falling telephone poles and flying highway signs heralding the demise of Prohibition with ads of 3.2% beer.
>
> When the car engine drowned out we wrapped it in a baby bed sheet, then found the roads back into Corpus Christi closed. We spent the night on car-seat beds on the floor of the Alice Bank and Trust Company, trying to dry birds-eye diapers by hanging them over the tellers' cages.

Also knowledgeable about hurricanes is a member of the younger generation, Kim Conner, who was a student at Austin Community College when she gave me this recollection of her experience in Hurricane Alicia of 1983.

> Hurricanes are a part of life in Galveston, like snow in the north. We board up houses, stock up on food, and buy candles. . . .
>
> Not so [some] newcomers. They panic and pack up and leave. . . . Houston [television stations] encourage it. . . .

STORMY

After boarding up the houses [for Hurricane Alicia] we—my mom, dad, my friend Yvonne, and I—went to the seawall to watch the waves. I wasn't impressed. They had been much bigger with Hurricane Allen—and he hit three hundred miles south.

We went home to fix supper. Outside it looked as if we might finally have a storm. The wind was about 50 m.p.h. and the rain was hard. During supper the storm boards on the side of our house blew off. My dad decided to put them back. We couldn't persuade him otherwise, so mom and I went out to hold the ladder.

He almost had a board back on when a gust of wind picked it up and picked him up, too. The board landed about fifteen feet away. Dad landed first on me, then on my car, and then on the concrete. We [gave up on this board].

About midnight the power went off. Eventually no part of the city had electricity except where emergency-produced. But our phones stayed on, and we never lost water pressure, though it was very low. No flushing the toilet. Around one o'clock we heard [on battery-powered radio] Radio Station KGBC announce they had to go off the air to switch to emergency generators. It was the only time during the whole storm they were off. When rival Radio Station KILE lost power, all its staff immediately went to KGBC and helped there until after the storm had passed.

Two o'clock in the morning. I notice rain is being pushed under the doors by the wind. Towels, please! We used up over twenty towels and that was by alternating them!

Three o'clock. Let's try to sleep. I go to my room, Yvonne to my brother's (he had a summer job in the Panhandle), my parents to theirs. Within half an hour Yvonne was with me. A storm board in her room had been knocked off, and she was afraid the window would break.

Things weren't much better in my room. The board was still intact, but it rattled, and wind whistled through it. We could hear tons of whatever was being hurled down the street. It was not conducive to sleep.

We went downstairs at four and spent the rest of the night changing towels and listening to the radio.

Daylight. We peek through a window at what damage was visible. We could see the sky through a neighbor's roof. Then to sleep again, or try to.

About seven some neighbors woke us. They, along with other neighbors, had gone out in the rain and wind (still 30 m.p.h.), checking out damage. We found out then that a tornado had hit houses across the street, next door, and behind us. We had never heard the tornado.

Considering the influence of hurricanes upon the Texas Gulf Coast it seems appropriate that the first airplane ever to fly into one of those tropical storms, showing the way for later hurricane-hunting flights, took off from Bryan Air Force Base, Texas, on July 27, 1943 (as narrated in "Into the Eye," by SMSgt. Vickie M. Graham, *Airman,* December 1984).

The flight began on a bet. Air Force Base personnel had heard of a hurricane heading toward Galveston, and some Royal Air Force pilots stationed at the base for training sarcastically questioned the ability of an American aircraft then in use, the AT-6, to withstand such turbulent weather.

Major Joe Duckworth, base commander, took up the challenge and recruited a reluctant second lieutenant, Ralph O'Hair, to accompany him as navigator. The loser would buy one round of drinks, silently assuming the loser wasn't Duckworth, along with O'Hair, in a storm-downed AT-6.

Forty minutes after taking off from Bryan, and after flying through storm turbulence never before deliberately experienced, they reached the eye of the hurricane. They circled it twice before flying back. O'Hair, who later admitted to being scared during the entire flight, still remembers the noise and the mass of pounding rain, the solid blackness, and the terrible turbulence.

Even stormier than Gulf hurricanes are Texas tornadoes, and the two often go together. Tornadoes usually cause less property damage (if more deaths) because they strike a smaller area. But tornado dangers are even greater than hurricane hazards, usually, because the winds are two or three times as violent. Tornadoes can occur anywhere in Texas—and, in fact, anywhere in the world—but since 1950 Harris County (Houston) has had more confirmed reports of those storms than any other county in Texas or the United States.

I have lived through hurricanes, but I have never experienced, firsthand, a tornado, and I hope that status quo holds. As I have mentioned, though, tornadoes have flitted close by me, and I remember one with great clarity. I wrote about it in *The Tornado* (New York: Doubleday, 1977). If my wish at the time had been granted, one of the most notorious "killer tornadoes" in Texas history would have swooped me up with the ease, but probably not the gentleness, of that storm we all know about from *The Wizard of Oz.*

The year was 1953, one of those dry times in Texas. I was employed as telegraph editor of the *Temple Daily Telegram,* but on Monday, May 11, I had the day off. Instead of laboring in the salt mines of journalism I was working much harder and more uncomfortably on my father's ranch between Temple and Waco, helping him water twenty or so head of cattle that had thus far escaped the fate doled out by an auctioneer at a livestock sale in Temple. The drought had, of course, forced that sellout of cattle.

The rock-bottomed creek that usually trickled into late-summer months before drying up completely afforded no help. An auxiliary water supply, the windmill standing near the same creek, had been given such hard use in recent months that the deep, spring-fed well from which it drew water was running perilously low. More immediately unpleasant for us was the knowledge that something had gone wrong with the reciprocating vertical rod used to transmit power from the wind-driven wheel above to pump below, so that the mechanism would not work properly. To get water for the bony cattle now hiding from raw sunshine in dust-powdered thickets somewhere around us, we had to disconnect the vertical rod at a point about two feet above the pump, then do the wind's job with gloved hands. Working single stints for as long as we were able, we would bend over the now-shortened rod and force up-and-down motions on it, quickly exhausting ourselves. The hard work caused water to spill from a pipe into a huge, circular concrete trough in a stream that impressed us as being very small for all the agony that went into the chore.

The labor required frequent breathing spells. One of us would pause from his work for a rest, sweating and burning under the midday sun, while the other would take over until he had used up, for the moment, his own arms, shoulders, back, and lungs. Despite this torture, there came occasional relief from the stinging sunshine when a lone cumulus cloud would place itself between us and our fiery persecutor long enough to cast a delicious shadow and make us wish for more.

So there *was* enough moisture somewhere to form clouds. Even as we worked the pump, they began to increase in size and number. We could see that, but what we didn't see or know about at the time was that a cold front had been creeping our way from the west, while a high-pressure system to the east was causing moisture to blow clockwise into our area on a southeasterly wind that came from the Gulf. I don't remember feeling much wind that day, and I didn't know the details of those meteorological occurrences until later, when I read "The Weather and Circulation of May 1953," by Jay S. Winston in the *Monthly Weather Review* for May 1953.

About to take place in our very own part of Texas was a collision of cold, dry air and warm, moist air—a happening that can result in stormy weather. But any storms certainly seemed far away as we called the cattle out of distant, dusty thickets where they had hidden from the blazing sun. Not all of them heard our beckoning shouts of "s-s-w-o-o-o-o! s-s-w-o-o-o-o!" So we set out, walking, to find the others, a task that required another hot half-hour or so. After that we tarried at the ranch to check out weak points in the fence.

The anvil-like top of a giant thunderstorm to the west sneaked up on us while we worked in a heavily wooded area. If any distant rumbles of thunder reached us, we must have attributed them subconsciously to the firing of big guns at Fort Hood, miles to the southwest.

We walked to the car and waited there to watch developments. The cloud grew closer and blacker and darkened a vast expanse of the western horizon, but it also showed a noticeable movement northward. Silently we urged it toward us and the parched ranch, even after noticing a green tinge at the distant base. *That* meant hail.

The thundercloud did indeed lumber on northward, but before long some trailing rainshowers dampened the ranch, then washed the highway as we drove southward toward Temple and my parents' residence.

A short time after I arrived there I had a telephone call from the managing editor of the newspaper where I worked. Could I drive to Waco, he asked, and investigate reports of a tornado that had struck the city? I did, and with awe saw the desolation wrought by the very cloud we had been courting only hours earlier. It had made a shambles of downtown Waco, demolishing brick buildings, smashing automobiles almost flat under tons of debris, and killing at least 114 persons. Never since that day have I yearned with quite the same ardor for a drought-breaking rainstorm.

Another individual who must be a bit nervous now when tornadoes are skipping around the vicinity is Lavonne Ishman, who has also given me a Texas weather reminiscence. She and her family and her sister's family were bound for California from Wichita Falls in June 1960, when they all unknowingly blundered into a tornado in the Lamb County community of Earth, on U.S. Highway 70. Several miles before reaching the place, they decided to stop there for gas.

> The sun was shining brightly, but one area of sky in the distance continued to hold our attention as we tried to figure out what gave it a peculiar color. Nearing the town, we began to see dust devils along the road. They became progressively larger until, suddenly, we were engulfed in swirling

sheets of red sand! Bob [Lavonne's husband] had to turn on the headlights to see, but we could hardly make out the car in front of us [driven by Lavonne's brother-in-law, Don]. At Earth we pulled into the first station we came to. Around us, cars began speeding every which way. Sirens were screaming. Trees were swaying mightily. Don and Bob jumped out and ran into the station to find out what was going on. The station manager was stuffing his money into a bag. He looked up and yelled, "Get out of here, you fools! There's a tornado! Get out!"

Not knowing where to go, our drivers headed down a residential street. When they saw an older couple walking toward a storm cellar they pulled over and stopped. The old gentleman, seeing the fright on our faces, hollered, "Come on! Come on! Get in here with my wife and me!" The seven of us, running and falling in the powerful winds and the rain, finally made it to the underground shelter.

We still think of that as the day when Earth rose up to meet us.

Our daughter Mary has also experienced at least one close call with a tornado. Always an independent person, she rented a mobile home in a rural area near Temple and moved into it, alone, in the spring of 1976. She had just turned nineteen.

Independence, I decided, felt good. One Saturday I was sitting under a tree in my back yard drinking a tall glass of tea and watching my ten dogs and countless cats melt in the unseasonable heat. Even the horses in the back lot seemed awfully hot.

About three o'clock I noticed a bank of gray clouds. In a very short time they turned from gray to blue to black.

I retreated into the trailer, feeling growing anxiety. Before this, any time bad weather threatened I had the safety of a sturdy little brick house just five miles away in Temple, where I had lived with my grandparents for several months. The sky grew even darker. The humidity seemed suffocating, and the air was still. At the base of the cloud I could make out a green hue: hail.

Feeling less independent, I drove to a nearby coffee shop built on a hill overlooking western Temple. With hot coffee, a booth beside a big picture window, and people around, I waited for the storm.

While I watched it I saw the tail of a tornado form at the base of a cloud. Slowly, easily it dipped to the earth, only a short distance from where I had come.

I stood up in the coffee shop and told the twenty or so customers we had a tornado moving our way. We all watched the tornado move across an area, leaving behind destruction and a few deaths, as I found out later.

Then, effortlessly it seemed, the great tornado picked itself up and went back into the sky and into oblivion. But I didn't go back right away to the trailer home, though when I did I found it undamaged. Instead, I drove to visit my grandparents in their brick house. Independence could wait awhile.

A brick house couldn't ensure safety in a bad tornado, of course, but the old reliable tornado haven, the storm cellar, doesn't seem so common in Texas as it was when I was young. My grandfather's farm had one near the house, I've heard, but it had disappeared, probably the result of a cave-in, before I came along. I have heard, also, that my weather-watching grandfather never went into it; he simply built it for others.

Another of my guest weather chroniclers, Barbara Waldrop, now of Austin, wrote about storm cellars in the Panhandle, where she grew up—at Borger. "Do you know where I can find a hole in the ground?" she began in a story given to me.

I'm serious; I need that hole—for my security blanket, if you will.

This obsession goes back to childhood. My home as a child was in the Panhandle of Texas—tornado country. Now I live in Austin. There is a conflict. I cannot be convinced Austin isn't tornado country, too. A storm cellar, my hole in the ground, is a wonderful haven of safety during severe weather. Austin does not have a good supply. The ground in my back yard is rock covered by two inches of soil. A neighborhood storm cellar just can't be built in solid rock.

In Amarillo, as I remember, a severe weather alert can unite the neighbors in underground concrete rooms, often furnished with a mattress and lawn chairs. Pickaxes, shovels, and lanterns come as standard equipment. Someone always brings a radio.

People hurry, shaking and frightened, to the cellars as tornado warnings go out. Women complain to their husbands: "John, get in here. It's too late to look for the tornado. You don't have time."

"Hmm. It's a false alarm," comes the reply. "I don't know why you are so scared."

The men seldom admit they are frightened.

Many more trips are made to the cellar than are needed. Some people sleep every night in the cellar during stormy season. Lives have been saved by such habits. Friendships are formed as neighbors share their shelters. People sometimes share expenses, too—two or three families build a cellar together.

Regular trips to these shelters are common occurrences in most other Panhandle towns. My home town, Borger, is an oil refinery town. Phillips Petroleum Company buried gasoline storage tanks there, and these tanks were used as community storm shelters. But I recall going there only once as a child.

My grandmother's refuge was a root cellar: a hole dug in the ground with boards and piled-up soil for a roof. Grandmother stored her canned goods there.

Not only do I want a hole in the ground because of a respect for Texas storms, but I'm not acquainted with my neighbors. How can I possibly get to know them if we don't spend a night or two below ground, the way I remember from the Panhandle?

This book is nearing an end, and it is still skimpy on some weather phenomena like fog, dew, and frost, and on some other storms like lightning and hail. Any seeming slight is unintentional—but any reader who has lived in Texas for very long can fill in blanks from his or her own experience.

Lightning has my greatest respect. I acquired this from my mother, who never could sit quietly during lightning after she saw a fireball traipsing through her house, eventually without doing any damage, in some storm long ago.

Our current residence has been struck once by lightning—in 1968—but it proved beneficial. The lightning struck and tore down a television antenna that a previous owner had put up. For months I had been intending to take it down but had never been able to psyche myself up for climbing on our sloping roof to do it. Insurance paid for replacement of some shingles that were also destroyed, and for some wiring repairs inside. Only four nights ago (as I write this), the lightning came back, perhaps to remind me to retain a degree of humility or simply to prove it *can* strike again in the same place, or at least close to it. On that night, a Thursday, lightning struck a telephone pole across the street from us, throwing an eight-foot piece of shattered wood into our front yard not far from a window, and leaving long splinters stuck in the grass like so many Indian arrows.

Our son-in-law, David Ballew (Mary's husband), was not greatly impressed when I told him that story just yesterday. He has had his own experiences with lightning. When he was ten or twelve years old, he said, he was walking through a field with his father, looking for Indian arrowheads that still abound there, when a sudden crash of lightning struck between them. David thought his jaw had been knocked out of joint, and his father felt as if he, himself, had lost a shoulder; but amazingly, they suffered no permanent injuries. He had other stories to tell, such as the time when lightning struck his parents' residence and removed from a four-by-five-foot mirror a piece resembling a half-moon without leaving an edge sharp enough to cut human skin.

I stopped telling him my lightning anecdotes. Nor would I recite them for meteorologists with the National Oceanic and Atmospheric Administration, because in their research they have probably heard almost everything—and certainly enough to call lightning "the underrated killer." Between 1940 and 1973, they have found, an average of 204 Americans were killed each year by lightning, compared to 132 for tornadoes, 90 for floods, and 54 for hurricanes.

When lightning begins flashing and thunder crashing, I have an affinity for going inside a house or building, whether I am in Texas or somewhere else. Lightning bolts from one-half to one inch in diameter can contain from 100 million to 1 billion volts of electricity, and the peak temperature can rise to 50,000 degrees Fahrenheit—four times as hot as the sun. Whoever it was who measured those things had a greater curiosity about lightning than I've ever had.

Hail can be another monster, as we all know. I have been caught in hailstorms while driving along Texas streets; stones approaching baseball size would bang into the car top with a sharp noise almost as uncomfortable as that made by three-inch guns on a warship. What kept the hail from crashing through the car top remains a mystery; the metal didn't seem that sturdy.

While driving through one such hailstorm I reflected that the din must be similar to a type of torture I read about long ago in *Darkness at Noon,* I think it was—torture that left not a mark on the human body. The captors placed a metal bucket over the victim's head and banged on it with large sticks until they got a confession or whatever it was they were seeking.

But *hail* certainly can leave marks. Leslie Carpenter, another weather chronicler for this book, recalls a double whammy dealt by Texas weather to the family farm near Bertram. After searing drought, rains finally came, and wheat grew. Then a hailstorm struck, destroying everything and causing $30,000 in damage. For the first time in ten years, they had not taken out crop insurance.

I'd prefer to complete my hailstorm coverage on a happier, if unrealistic, note, and that brings me to the story of the world's largest hailstone. It fell on Waco years ago.

The story, apparently once well known in the city, has been largely lost in the mists of human memory, but it still surfaces occasionally, retold with varying details. Most recently it was recounted in a column written several years ago by the late Dallas journalist Frank Tolbert, then, following that, by Wacoan Leo Lyons in the April 23, 1982, *Waco Tribune-Herald.* With Lyons's permission I'll quote him.

Several years ago . . . Tolbert recalled an event that in 1920 produced what came to be known as "The Giant Waco Hailstone." The famed Robert L. Ripley reported it in his "Believe It or Not" syndicated newspaper cartoon, declaring that a 10-pound hailstone, the largest ever, fell on a street in front of a downtown Waco hotel before about a dozen or more sober witnesses taking shelter in the hotel's entrance.

So the story goes, one John Fitzhugh, a railroad executive, who was known as "Uncle Johnny" to his friends, was a guest in the hotel that day. His room on the second floor was situated where he could see the people standing under the portico.

Those were Prohibition days and Fitzhugh was having a few illegal cocktails in his room. While sipping along, he chipped away at the big square of ice that the bellboy had brought up in a silver bucket, and the ice soon took the shape of a globe. About that time it started to hail—small pellets (probably pea-sized). Suddenly, forces conspired to inspire "Uncle Johnny," who had a bent for mischief. He carried the chunk of ice into the bathroom and held it under hot water until its surfaces smoothed out into an almost perfect ball. At the height of the hailstorm, he tossed the ice into the [slushy, hail-covered] street right in front of the hotel entrance. Cries went up and anxious faces peered skyward. A doorman rushed out, picked up the giant "hailstone" and carried it into the hotel, where the ice was weighed on kitchen scales at 10 pounds.

When a reporter and photographer arrived, Fitzhugh decided things had gone far enough. He called to the reporter, a friend, and told him it was all a joke, that the hailstone was right out of the ice plant. That irked the reporter, who knew he had a good story. He advised Fitzhugh he had half a dozen witnesses who had seen the thing fall right out of the sky.

There was no dissuading the reporter, and the next day there were headlines complete with verification from the doorman and other awe-struck witnesses.

Now I believe I know why my grandfather, that long-suffering Bell County farmer, generally refused to record on the walls of his old barn weather notations copied from the local newspapers or anywhere else. He wasn't formally educated, but he was a questioning man who spotted hoax and hypocrisy easily, though quietly, and he had infinite confidence and self-reliance despite his humble station on earth.

With such a thin line dividing accuracy and error, truth and falsehood, he accepted only what he had detected with his own senses—or, in the case of those weather observations, only what he had seen for himself; and he had a frontiersman's eyes, clear and sharp.

He was a genuine weather-watcher, in those days before watching weather meant turning on the local television news. My grandfather was never profane about the weather, but weathercasters might have been hard for him to endure in silence.

Right now I am rereading the last post card he ever scrawled and mailed to my father: "Will Go to Temple This Evening to stay for a while [with a daughter]. Am getting better. Glad to hear from you. Write soon." That message tells me, these forty years afterward, that he was *not* getting better. There is no mention at all of the current state of the weather, and that was one item he had never failed to include. Still, he hadn't lost interest, as soon became evident. He had lost only an ability to observe weather for extended periods the way he had always done.

I was several thousand miles away then, in the Pacific, but I heard later that some of my grandfather's last questions concerned the exact state of the weather outside. It was snowing, they told him at one point, and I don't doubt that he enjoyed a walk through it, in his imagination, before he died.

The changing seasons seemed never to throw him out of step with himself or, really, with anyone else—not even the sometimes odd, stormy seasons the Lone Star State can have. He was a native Texan in every sense, and I believe he loved them all.

SOURCES

The text of this book is based to a large extent on my personal experiences, observations, and collection of weather stories dating back to my first book, researched in 1956, and in a few cases even earlier than that. Sources of most of the rest of the information are specifically identified in the text.

Archival sources used in this book came from the Rosenberg Library, Galveston; Texas Collection, Baylor University; University of Texas Archives and Newspaper Collection; and Waco Public Library. I also used three of my own earlier books concerning weather: *A Weekend in September* (Texas A&M University Press, 1980; reprint of 1957 edition); *Growing Up in Texas,* which included other contributors (Austin: Encino Press, 1972); and *The Tornado* (New York: Doubleday, 1977).

Some books not mentioned in the text but of varying weather interest are the following:

The Great Plains, by Walter Prescott Webb (Boston: Ginn, 1931).

The Handbook of Texas, three volumes (Austin: Texas State Historical Association, 1952, 1976).

The Texas Almanac, 1984–1985, edited by Michael T. Kingston and Ruth Harris (Dallas: A. H. Belo Corporation, 1983).

Texas Weather, by Jud Ashmore (Oklahoma City: England and May, 1975).

Texas Weather, by George W. Bomar (Austin: University of Texas Press, 1983). This is what I consider the standard reference book on the subject, written by a meteorologist.

Tornadoes over Texas, by Harry Estill Moore (Austin: University of Texas Press, 1958).

Weather, by Paul E. Lehr, R. Will Burnett, and Herbert S. Zim (New York: Golden Press, 1975). For young readers and for adults who know nothing about meteorology, this is an excellent book.

The World Almanac & Book of Facts, 1985 (New York: Newspaper Enterprise Association, 1984).

Newspaper information most frequently referred to came from the Associated Press, *Austin American-Statesman, Dallas Morning News, Dallas Times Herald, Fort Worth Star-Telegram, Houston Chronicle, Houston Post, New York Times, San Antonio Express, Temple Daily Telegram,* and *Waco Tribune Herald.*

INDEX

NOTE: All place names are for cities and counties in Texas, unless otherwise indicated.

INDEX

INDEX

INDEX

Photo Credits

Pages 9, 91:
Austin History Center, Austin Public Library

Page 87:
Charles C. Raines, Chalberg Collection, Austin History Center, Austin Public Library.

Pages 4, 59:
Photograph Collection, Barker Texas History Center, University of Texas at Austin.

Pages 62, 67:
Vandale Collection, Barker Texas History Center, University of Texas at Austin.

Page 77:
La Retama Public Library, Corpus Christi, Texas.

Pages 41, 44, 51, and 100:
Institute of Texan Cultures, San Antonio, Texas.

Pages 16, 22, 25, 29, and 34:
Courtesy of the National Oceanic Atmospheric Administration, National Weather Service, San Antonio, Texas.

Page 97:
Leon Hooten, Wichita Falls Times and Record News.